Universal Dynamics and Big Bangs

www.novapublishers.com

Advances in Astronomy and Astrophysics

More information about this series can be found at
https://novapublishers.com/product-category/series/advances-in-astronomy-and-astrophysics/

Physics Research and Technology

More information about this series can be found at
https://novapublishers.com/product-category/series/physics-research-and-technology/

Bent Rolf Pettersen

Universal Dynamics and Big Bangs

DOI: https://doi.org/10.52305/OHBP4354

NOTICE TO THE READER

Library of Congress Cataloging-in-Publication Data

ISBN: 979-8-89530-746-5 (Softcover)
ISBN: 979-8-89530-872-1 (eBook)

Published by Nova Science Publishers, Inc. † New York

Contents

List of Illustrations

Preface

Correct physics must work with quantum mechanics and astrophysics. "A law is no law if it only applies in some cases" (Stephen Hawking).

This book is an attempt to show a model of physics which works within all fields. In order to do this, we have to take a deeper look at universal energies, their properties and how different energies react with each other. Based on this we can establish a model which explains how these energies can be concentrated and organized in organized energy systems, like we see in the elements in the periodic table of elements.

This model can explain the fundamental forces, how to build and dissolve organized energy systems, and also explain universal dynamics like dark energy, dark matter, gravity (the attractive force). It can also explain the dynamics of stars and black holes, and offer an explanation of Big Bang.

The Ilefos model is an alternative conceptual qualitative model of energies which offers a new approach to physics, which seems to work with all sides of physics.

Chapter 1

Introduction

The universe is unimaginably vast. The visible universe is only a small part of our universe. Our universe consists of various forms of energy, organized into a system governed by universal physical laws.

Our universe is one of many universes. Each universe is governed by a universal law. The energies and governing laws may vary across different universes, but they should have universal energies that allow communication between universes.

Energies in our universe can be concentrated and organized into quarks, particles, matter, and universal phenomena.

In this book, I will explain the nature of energies and how they interact and form matter. The book will also explain stars, black holes, and galaxies.

A Big Bang is an immense release of energy. The book will explain how a Big Bang can be formed, and how the Big Bang is a recurring event according to the Ilefos model.

The energy balance of the universe will be explained. To achieve this, we must explain gravity as well as both material and dimensional forms of energies. Universal physics must work in all fields, including the quantum field and astrophysics, and explain the dynamics of the non-visible universe. Dark energy, dark matter, universal energy fields, cosmic dynamics, and domain walls are among the key topics explored in this book.

This book builds upon the Ilefos model, which explains energies, gravity, and attraction, and works on quantum mechanics, astrophysics, and universal physics. This is a qualitative alternative conceptual model of gravity and energies.

> Different energies have different properties. These properties determine how the energy reacts to similar or different energies.
>
> A force arises from interactions between energies. Energies generate forces based on their inherent properties mutual interactions.
>
> Bent Rolf Pettersen, 2024

Physics is the study of energy and the interaction between energies. Interaction between energies is the story of the universe.

Dimensions are areas dominated by dimensional energies with their own physical laws. Material dimension is one of several dimensions. Dimensions are essential for understanding existence.

Dimensions interact with each other. There are points of contact between dimensions. Through these, interaction between dimensions can take place. This is very important knowledge about nature and the universe. Interaction between different dimensions takes place through special universal energies. Understand these energies, and you will understand everything.

Bent Rolf Pettersen, 2023.

The Universe is energy -and the organization of that energy.

Bent Rolf Pettersen, 2022.

When a massive black hole achieves critical mass and can't hold the mass together, all energies are released, and we have a Big Bang.

Bent Rolf Pettersen, 2021.

Chapter 2

The Universe

Our visible universe is just part of the grand universe. The universe consists of the visible universe, which we can observe, and the universe beyond. Our visible universe originated from a Big Bang that released immense energy and gave rise to the structures we can observe today.

The visible universe is limited to what can be observed and is limited by the speed of light. Beyond the visible realm lies what may be termed the broader universe — including regions predating or existing outside the scope of our Big Bang.

When the term "universe" is used, it refers to both the visible universe and that which lies outside the visible universe, the grand universe.

Our energies and our laws of physics define our universe. Other universes may consist of other energies and other laws of physics. Our universe is one of countless universes in the multiverse.

I will try to explain how our visible universe, which started more than 13 billion years ago in a Big Bang, was not the beginning of the universe. Big Bangs are recurring events in the universe. Each universe seeks an energetic equilibrium, and Big Bangs function as redistributive events that restore or restructure this balance.

To explore this perspective, I introduce a universal physics grounded in the interplay between material and dimensional energies. This physics has to work in all fields to be universal, in both the quantum field and astrophysics. It has to explain dark energy, the accelerating expansion of the visible universe, and dark matter.

I start by explaining how one attractive force can explain atomic binding, gravity, and other attractions in the universe. The dynamics of quarks, particles, and matter will be presented by the Ilefos model. The same model will also be used to explain large-scale phenomena such as stars and black holes. All this will be in accordance with the Big Bang and a universe in energy balance.

Chapter 3

Energies

Energies give rise to forces. Understanding the nature of an energy is essential to understanding how forces operate. Each type of energy possesses distinct properties, which characterize it.

A force emerges as the result of interaction between different types of energy.

Energies create forces based on their properties and the interaction with each other.

To understand the universe, we must understand the physics of nature's smallest components. The energetic principles that govern these parts also apply universally.

Therefore, we begin by exploring how energies can organize and concentrate into quarks, particles, and matter. This foundation enables us to address larger universal phenomena—such as stars, black holes, the nature of so-called "empty" space, and dark energy.

By understanding these mechanisms, we can grasp the broader dynamics of the universe, including the Big Bang and the observed accelerated expansion of the visible universe—appearing faster than the speed of light.

Finally, we must address dimensional energies, which operate within and around material energy clusters and exert significant influence on both matter and the cosmos.

Atomic quarks can accumulate and store large amounts of diverse energy types. Some specialized quarks can even convert energy through a process known as Atomic Phase Displacement.

In the following sections, I will explain the fundamental forces of the material world and the energies that generate them —framed through the Ilefos model, which integrates quantum field theory and astrophysics.

Chapter 4

The Universal Energy System – The Rules of Physics

The universal energy system forms the foundation of all physics. It defines the rules for energies —their behavior, properties, and interactions. This system constitutes the universal order.

Energies can be coordinated and converted. For instance, energy transformations occur in matter and atoms, such as in electromagnets and generators.

Energies can be gathered, concentrated, and organized in energy systems, as we see in quarks. Quarks can assemble into structured formations, observable in both photons and atoms. The rules for how energies build these structures and the dynamics between them are also part of the universal energy system.

The universal energy system governs the dynamics of the universe. Each universe possesses its own energy system and corresponding physical laws. Our physics applies to our universe, while other universes have their energies and regulations.

In our universe, two primary energy systems operate: material energies and dimensional energies. Their respective properties and behavioral principles are referred to here as physics and dimensional physics. Universal physics rules all energies in the universe, both material and dimensional.

In this book, we will focus on material energies. A separate chapter later in this book explores dimensional energies; this chapter focuses exclusively on material energies.

Different energies have different properties. A force is a reaction between energies. The physics in our universe is based on our universal energies and how they interact, both as individual energies and in organized energy structures.

These foundational distinctions lay the groundwork for understanding how energies manifest, interact, and structure the universe in both physical and pre-physical dimensions.

Chapter 5

Gravity

Gravity is the attractive force that pulls atoms, matter, and universal phenomena (stars/black holes) towards each other.

Albert Einstein explained gravity in general relativity as the curvature of spacetime. General relativity describes gravity as the warping of a hypothetical spacetime fabric—an abstraction that lacks a physical explanation. It gives good mathematical calculations of observations in the universe of the attractive force (gravity). However, since the nature of spacetime remains unexplained, I propose a different approach.

Isaac Newton's law of universal gravitation describes gravity as a force causing two bodies to be attracted to each other. This attraction is defined as the product of their masses, inversely proportional to the square of the distance between them. This classical description of gravity works for most applications. It is not as accurate as general relativity but is a reasonable gravity calculation from 1687.

Isaac Newton's law of universal gravitation:

$$F = G\frac{m_1 m_2}{r^2}$$

The force F determines the attraction between two masses (m1 and m2). The force is the gravitational constant (G) multiplied by mass 1 and mass 2 divided by their distance squared.

This formula shows that the attractive force between masses becomes weaker with the distance between them.

By enhancing this formula through a deeper understanding of the attractive force, we may arrive at a more accurate explanation of gravity. The following sections explain gravity and the attractive force according to the Ilefos model, which integrates quantum and astrophysical domains. This model refines Newton's law of universal gravitation to account for underlying energetic mechanisms.

Chapter 6

The Attractive Force

The attractive force pulls gravity-affected objects toward each other. Gravity affected objects are objects with mass, like particles, matter, and universal phenomena such as stars and black holes. They are all affected by the pull of the attractive force.

The attractive force can create bonds between masses. Atoms can have strong atomic bonds, as observed in metals and solid matter. These bindings form close to atoms where the attractive force seems strong. At a distance, the attractive force forms a weaker attraction, gravity. How can the same force account for both atomic bonds and gravity? Could it also explain the dynamics of the visible universe?

In order to have one attractive force explaining both behavior of the quantum world and astrophysics, we have to explain the nature of the attractive force, how it decreases over distance, and how it can work over great distances.

The Ilefos model explains the attractive force as a special energy, ilefos. Pulses of ilefos, ilefos energy units, have a strong attraction to each other and other ilefos energy. When the energy units are released from their source, their attraction links them together and forms a track of ilefos energy units, a gravity track. The source of an ilefos (gravity) track can be matter, dark matter, stars, or black holes.

When two ilefos tracks meet, they create a bond. The strength of the weakest track determines the strength of this bond. If two strong ilefos tracks meet close to an atom, they create a strong connection, an atomic binding. If a weaker ilefos track meets an external track, they form a weaker connection, gravity.

An atom has strong enough ilefos tracks to create atomic bindings close to the nucleus. The gravity track is strong and close to the source, but it weakens with distance from the source. If ilefos energy units retain their strength over distance, then the track itself must be dynamic in nature.

Chapter 7

The Nature of Gravity Tracks

The nature of the ilefos energy units must include mutual attraction, both between the units themselves and toward the strong nuclear force.

The strong nuclear force holds the quarks and the atomic elements together in a structure. This requires the presence of a quark that contains accumulated ilefos energy—here referred to as the plus quark I call this quark plus quark. In order to keep this energy and quark(s) together in the nucleus, the ilefos energy has to be attracted to the strong nuclear force of the nucleus. The plus quark (ilefos quark) may correspond to the role of the Higgs boson, which in this framework is interpreted as an energy-condensing resonance structure.

If the ilefos energy units are to be released from the nucleus, they must overcome the attraction of the strong nuclear force. They must be released with a speed and force strong enough to surpass the strong nuclear force. When released from the nucleus, the strong force still affects the pulses of ilefos energy. This residual drag results in a curved trajectory, causing the ilefos track to adopt a Fibonacci-like spiral.

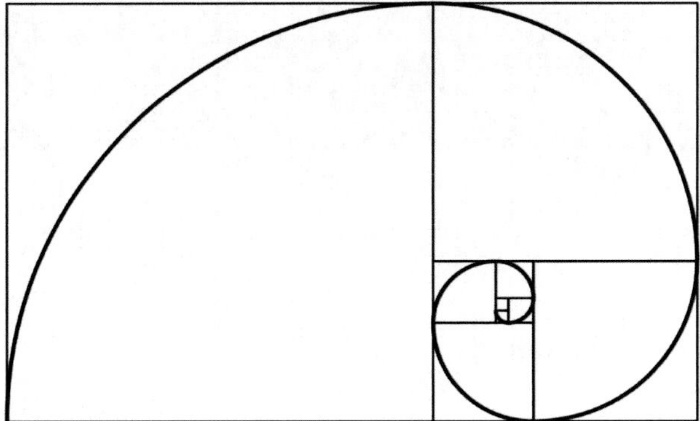

Figure 1. Fibonacci spiral (Dicklyon).

With distance from the nucleus, the drag from the strong force will be reduced. The ilefos energy units will then increase in speed. The ilefos track will slowly straighten, thus creating a Fibonacci-like track. With distance from the nucleus, the gravity track will be almost straight.

When the ilefos energy units' drag from the strong force diminishes, they will continue accelerating until they achieve their energy speed, their natural speed in the universe.

With accelerating speed, the distance between the ilefos energy units in the gravity track will become longer. The attraction between the ilefos energy units will be weaker with increased distance. The ilefos gravity track will thus become weaker/less dense with distance from the source.

Ultimately, the distance between the ilefos energy units becomes too long to hold together. The ilefos track then dissolves in an ilefos release. The ilefos energy units then become free ilefos energy units, with no connection to the source. Once released, these free ilefos energy units are interpreted as contributors to the field commonly referred to as dark energy.

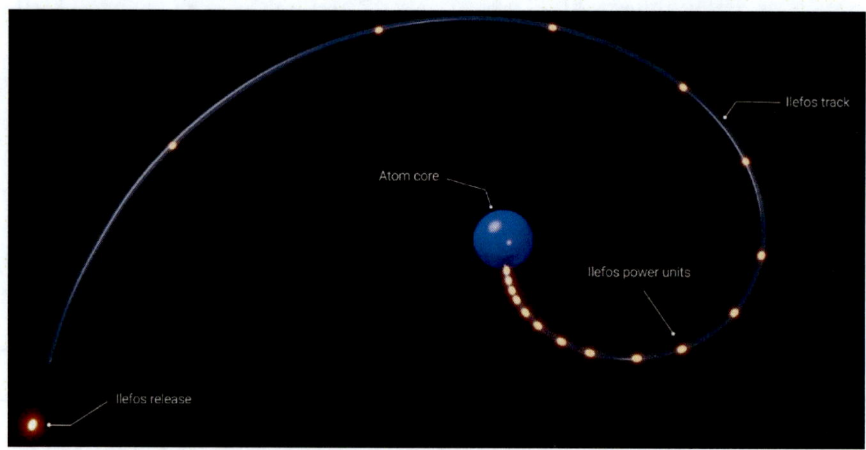

Figure 2. Pulses of ilefos energy units from an atom. They attract each other and form an ilefos track (gravity track). (Stig Boe/Mohammed Ismaiel).

How Do Ilefos Tracks Interact?

When two ilefos tracks meet, the energy units in the different tracks attract each other and create a bond. The ilefos energy units have a speed and "mass," making them continue their travel. The ilefos tracks split, but they do not exchange ilefos energy units.

With distance from their source, different ilefos tracks will have little chance of connecting if it were not for the attraction between the tracks. The ilefos energy in a track makes the track bend towards external tracks, which increases the chance of two ilefos tracks meeting.

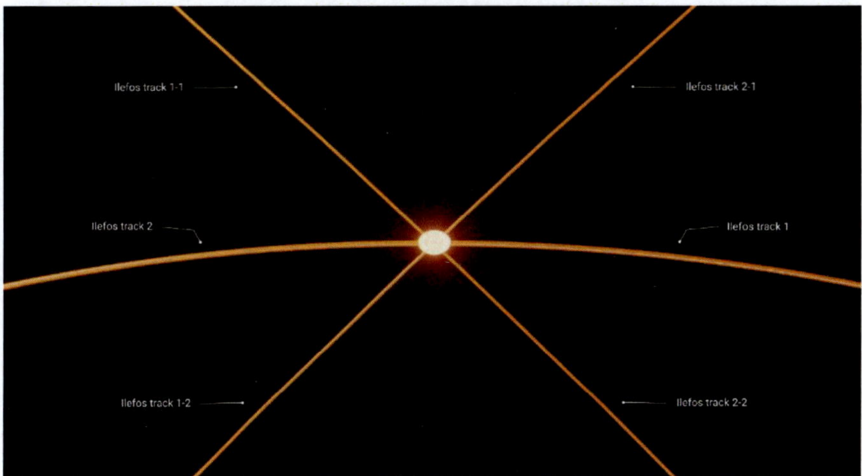

Figure 3. Two ilefos (gravity) tracks meet, connect, and split. (Stig Boe/Mohammed Ismaiel).

Stars and black holes release extreme amounts of ilefos energy units. The ilefos energy units form massive and dense ilefos tracks, which are much longer and stronger than the gravity tracks of matter. These tracks are very dense, and the ilefos energy units can hold together, due to mutual attraction, for a very long time. These gravity tracks, therefore, work over very long distances before they dissolve.

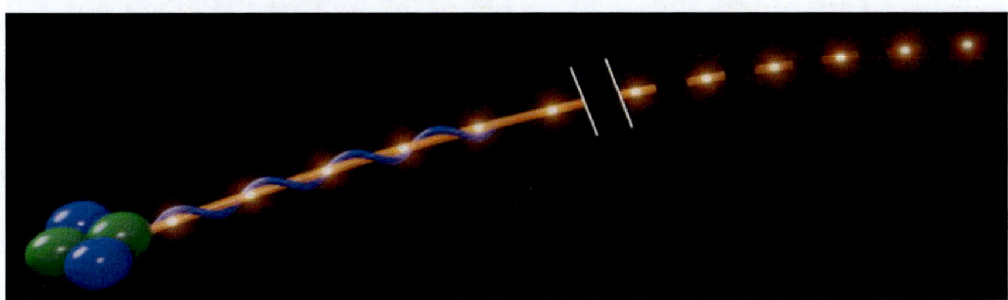

Figure 4. An ilefos track dissolves in ilefos release and into free Ilefos energy units (dark energy). (Stig Boe/Mohammed Ismaiel).

Chapter 8

The Repulsive Force

The attractive force pulls atoms toward each other. If strong enough, this attraction may lead to a collision between atomic nuclei, potentially crushing them.

The atoms have a repulsive force to prevent this and preserve the structure of the atoms. This force works only in the immediate area around an atom.

The repulsive force forms a shield around the atom. It must operate outside the nucleus but cannot be an intrinsic part of it. The nucleus consists of quarks and atomic elements held together by the strong nuclear force. Therefore, repulsive force cannot have a repulsive force towards these elements, repel them, and be a part of the nucleus.

How can the repulsive force be emitted outside the nucleus and repel external atoms? The repulsive force must be sent out from particles outside the nucleus. I therefore have to introduce a new particle and a new energy: Uni particle and TE energy. In this model, these particles conceptually replace the role traditionally ascribed to electrons in orbit around the nucleus.

In the Ilefos model, energies are shared between atoms in energy tracks, not through particles (electrons). Energy tracks are formed based on the energy (dielectricity) properties, which make the energy units form energy tracks. Read more of this in the chapter "Dielectricity."

Around the nucleus, we must have special particles in orbit. These particles, uni particles, have quarks of the repulsive force, TE quarks with TE energy, which have a repulsive force towards external atoms' uni particles. The uni particles also have to be attracted to the nucleus for the nucleus to hold them in orbit around the atom. The uni particles, therefore, must also have quarks that produce a strong nuclear force, which I call the YT quark. The uni particles are then attracted towards the nucleus, which holds them in stable orbits around the atom, and they have a strong repulsive force towards external uni particles.

The uni particles in orbit around the atom then create a repulsive shield against other atoms with uni particles. The uni particles prevent collisions between nuclei and help preserve the atomic structure.

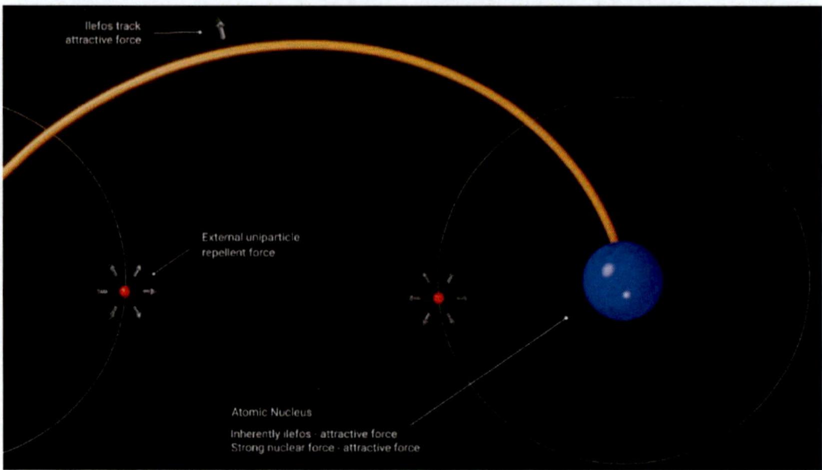

Figure 5. Uni particles' repulsive force help atoms keep distance between atoms (Credit: Stig Boe).

If two atoms are forced toward each other, they will increase the repulsive force of the uni particles. As we see in friction, this extra energy may be perceived as heat. If the uni particles are not able to stop the external atom, we will have a nuclear collision. This collision can knock an atomic element loose, which may dissolve and release its energy. In very strong collisions, the whole nucleus might be destroyed, knocking all atomic elements and quarks loose. These elements and quarks might dissolve and release immense energy. We see this in nuclear fission bombs and stars when matter dissolves and releases energy.

Figure 6. An atom with ilefos (gravity) tracks and uni particles (Credit: Stian Kongsvik).

Table. Quark Composition of a Uni Particle

TE quark	Responsible for storing and sending out TE energy, the repulsive energy.
YT quark	This quark stores and handles YT energy, the strong nuclear force. This energy holds the quarks together in a particle. The quark also has an attraction towards the strong nuclear force of the nucleus. This attraction holds the uni particle in an orbit around the nucleus.
Neutral quark	The neutral quark is responsible for converting energies in the uni particle. The quark can convert all energies, but primarily converts the strong nuclear force (YT energy) to the repulsive force of the uni particle (TE energy).
Alfa quark	Alfa quark organizes and runs the local quark group (uni particle) in close collaboration with the atomic quark of the atom.

Each neutron administers a uni particle. The number of uni particles is usually the same as the number of neutrons in the nucleus. The exception is normal hydrogen, which does not have a neutron. In this case the uni particle is run by the proton. This results in a weaker uni particle.

In isotopes, atoms with more neutrons than protons, the atom also produces extra uni particle(s) served by the extra neutron(s).

While the attractive force pulls atoms toward each other, uni particles orbit the atom to maintain structural integrity. These particles form a repulsive shield composed of TE energy, which pushes against the uni particles of neighboring atoms. This interaction maintains distance and helps prevent destructive nuclear collisions.

Chapter 9

Dielectricity

Dielectricity is the dominant form of energy used for exchange and conversion processes in ordinary matter.

Electricity is the energy term for the movement of electrons. In the Ilefos model, energy is transferred through energy tracks, which are tracks of energy pulses.

When houses to share energy, they do not exchange batteries ; they use transmission lines. We must assume that atoms also use the most effective method to exchange energy. Since energy in the Ilefos model is transferred through tracks of energy pulses, I call this energy dielectricity.

All atoms have dielectricity tracks if they have surplus energy. At very low temperatures, atoms do not form dielectric tracks.

Different atoms prefer different energies. The most common outgoing energies are ilefos (the attractive force) and dielectricity. The quark composition of the atom determines which energy the atom prefers to release.

If the atom has several plus quarks (ilefos quarks, associated with the attractive force), it prefers ilefos and will send out stronger ilefos tracks (attractive force) when it has abundant energy. Iron is an example of an atom with several plus quarks.

If the atom has several energy quarks (dielectricity quarks/gluons), it prefers dielectricity and will send out stronger dielectricity tracks when it has abundant energy. Copper is an example of an atom with several energy quarks.

Dielectricity is the easiest energy for normal matter to convert. Therefore, this is the most common energy for atoms to exchange. Electricity and heat are examples of dielectricity transferred by atoms.

When an atom receives strong dielectricity and has more plus quarks (ilefos quarks) than energy quarks, it prefers ilefos energy. The atom then converts dielectricity to ilefos, and sends ilefos out in powerful ilefos tracks (gravity tracks). When a powerful ilefos track meets an external ilefos track from an atom nearby with several plus quarks, they form a strong attraction. This is magnetism, which attracts magnetic material.

If iron receives strong dielectricity, it converts it to ilefos, the attractive force, producing strong ilefos tracks. This process underlies electromagnetism (as in the operation of an electromagnet).

When an atom receives strong ilefos and has more energy quarks (dielectricity quarks) than plus quarks (ilefos quarks), it prefers dielectricity. The atom then converts ilefos to dielectricity and sends dielectricity out in strong dielectricity tracks.

If copper receives pulses of strong ilefos (magnetism), it converts ilefos to dielectricity and produces strong dielectricity tracks. This is a generator.

Dielectricity Tracks

A pulse of dielectricity, a dielectric energy unit, is attracted to other dielectric energy units. When an atom releases dielectric energy units, its mutual attraction makes it form a dielectric energy track.

The attractive force strongly attracts other atoms and forms atomic bonds. Dielectric energy units are also attracted to ilefos. When released from an atom, this attraction makes the dielectric energy track follow the ilefos track. When the ilefos track connects with an external ilefos track and make an atomic binding, the two dielectricity tracks also connect. This connection allows the transfer of dielectricity. If one track has stronger dielectricity, it can transfer this to the weaker track, transporting dielectricity back to its source. This is how exchange of energy (dielectricity) is performed between atoms. How much energy (dielectricity) an atom can handle (receive and store) depends on the number of energy quarks in the atom.

When the atom has filled up its quarks and has maximal energy in its energy tracks, it must eliminate the surplus energy in other ways. It then has to produce and release energetic quarks, which might form photons (light).

The dielectric energy units in a dielectric track also accelerate (like the ilefos energy units in ilefos tracks). When the speed of the dielectric energy units increases, the distance between the units in the track becomes longer. The strength of the dielectricity track becomes weaker with distance in the same way as an ilefos (gravity) track.

When the gravity track becomes very weak, the weakened dielectricity track will also be unable to hold on to the ilefos track. The dielectricity track then breaks with the ilefos tracks. When this happens, the dielectric energy units have accelerated to a very high speed with a long distance between the units. The mutual attraction between the dielectric energy units now becomes too weak to hold together in a track. The dielectricity track dissolves into free dielectricity energy units. Free dielectric energy units are pulses of dielectricity that are not connected to a source. These free dielectric energy units—no longer bound in tracks—are hypothesized to constitute a significant portion of what is observed as dark energy.

Sources of ilefos (gravity) tracks also send out dielectricity tracks, with the exception of black holes. In black holes the ilefos concentration (gravity) is too strong, and they do not release dielectricity, with the exception of where the gravity is weakest. In the "top and bottom" of a black hole, the gravity is weaker, and dielectricity and other energies escape. This is Hawking radiation.

Chapter 10

The Strong Nuclear Force

The strong nuclear force has attraction to ilefos (the attractive force), dielectricity, and quark matter. Quark matter is the building material of quarks. Quarks concentrate a specific form of energy, perform organizing functions, and store structural information. All quarks are composed of the same energetic substrate, which is inherently attracted to the strong nuclear force. The strong nuclear force can then collect and shape quark assemblies. Groups of quarks are shaped according to the universal energy template, which we can also call the atomic DNA of matter.

Granted, in this context, DNA is a misleading term. DNA means deoxyribonucleic acid, which carries instructions for the development, function, and growth of organisms. However, in the absence of a better word, I use it in connection with instructions to the functions and development of matter.

Figure 7. The strong nuclear force is the glue that makes quarks hold together in a structure. Credit Ola Tandstad.

The universal energy template —referred to here as the the atomic DNA —provides the blueprint for assembling quarks and quark groups. The strong nuclear force is the glue that holds the quarks together in an organized system. Photons, as the smallest quark groupings, can evolve into larger and more organized structures—ultimately forming protons and neutrons, which can be organized in advanced atomic systems. The periodic table lists known atoms. Atoms can, in certain conditions, further develop into forms associated with dark matter. This will be explained later in the book.

A special energy, the YT energy, creates the strong nuclear force. YT energy is concentrated, stored, and managed by YT quarks. In order to have an advanced quark assembly, we must have at least one YT quark. Most larger quark groups in matter have more than one YT quark. The structural strength of the nucleus depends on the number of YT quarks and the YT energy of the atomic core.

The strong nuclear force shapes atoms and other quark groupings, and is essential to the architecture of matter.

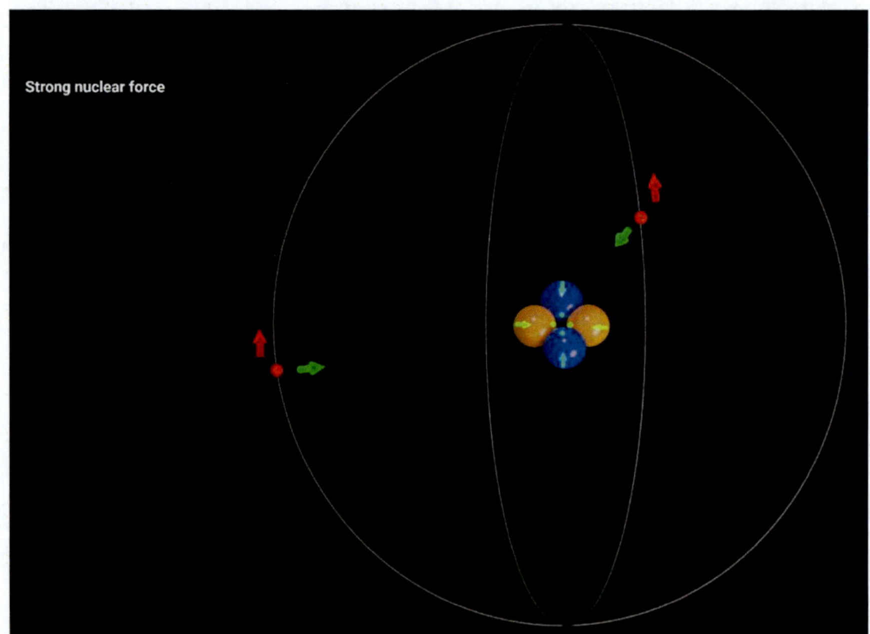

Figure 8. The strong nuclear force holds the nucleus together and the uni particles in orbit. (Credit: Stig Boe/Mohammed Ismaiel).

Chapter 11

Other Energies

The universe contains additional forms of energy beyond those previously described—namely, other material energies and dimensional energies.

Other Material Energies

The main energies organized in matter are called material energies. The dynamics of atoms are managed primarily by material energies. These energies also cause the dynamics of universal phenomena like stars and black holes. Other material energies are responsible for additional processes within matter that have not yet been detailed. These include forces managing the remaining atomic dynamics and will be addressed in the next book.

Dimensional Energies

Dimensional energies operate independently of the material energy system. They are organized within a dimensional energy system that exists both within and beyond the material dimension, yet do not directly influence it. Since dimensional energies do not interact with material, they are unaffected by drag forces such as those associated with ilefos. As a result, dimensional energies are not subject to speed limitations.

There are touching points between the material dimension and dimensional energies. In matter, we have special quarks centered around special dimensional energies. These energies are communication energies and energies connected with the atomic operating system.

A dimension should not be understood as a separate 'place', but as a structured energy system organized around specific dimensional energies. Non-material dimensions permeate and surround the universe, existing alongside and through the material dimension.

This book will focus on the material energies and material dimension. Dimensions and dimensional energies will be the focus of a future book.

Chapter 12

Dark Energy

Dark energy is energies that are not structured within matter or other organized energy systems. It consists of "free" pulses of energy. Dark energy consists of material energies, the energies we find in matter, and dimensional energies.

When matter dissolves in fission and atomic dissolution in stars and other places, the different quarks in matter dissolve and release the stored energies. Most of these energies do not form new quarks, but are released as pure pulses of different energies. These free, disconnected pulses of energy—without any link to source—constitute what we term dark energy.

According to the Lambda-CDM model (ΛCDM), which incorporates a cosmological constant (Λ) to represent dark energy, approximately 68% of the universe consists of dark energy. We must assume that all larger organized energy systems, like matter and stars, constantly absorb dark energy. This will be the theme for the later chapter, "Energy balance in atoms."

Dark energy pulses propagate at extremely high velocities, potentially exceeding the speed of light, due to their unstructured and frictionless nature. Photons, being structured and quantized units of electromagnetic energy, encounter greater resistance and thus travel at limited speed compared to unstructured energy pulses. Therefore, pure energy pulses travel much faster than organized energy. Read more of this in the chapter "Universal speeds."

Dark energy are pulses of pure energy. The composition of dark energy is all energies we find in matter and dimensional energies. "Empty space" is filled with free energy units, pulses of energy, called dark energy.

Chapter 13

Quarks

Energies can be concentrated within organized systems. Each system concentrates a particular type of energy. We can call an organization of a specific energy a quark In this book, a quark is defined as the organization and concentration of a specific energy.

To organize a specific energy, an organizing force is required—one capable of interacting with all quark energies. This energy must be able to interact with all energies in different quarks and is the energy behind the strong nuclear force. I call the energy behind the strong nuclear force YT energy.

YT energy can attract/interact with all material energies and tune into each quark's specific energy. YT energy forms a structure that can withhold and organize the particular energy of each quark.

In atoms and other organized energy systems, the quarks are organized by a special control quark. I call this management and control quark the atomic quark. The atomic quark is the "brain" or the central management system. A specific quark does not have an atomic quark, but an organized group of quarks needs one to manage the strong nuclear force, the quarks, and their energies.

In advanced organized energy systems like atoms, we also need quarks that perform specific atomic functions, such as converting and channeling energy in and out of the atom.

In a sub-organized system, like a proton or a neutron, there are two main types of quarks:

- Nuclear quarks
- Special quarks

Nuclear quarks are responsible for a special energy or a special task.

With each nuclear quark, there is a special quark. Each special quark carries the blueprint for its nuclear quark—it contains the structural 'software' or energetic "DNA" that maintains the quark's integrity. This quark contains the software and the "DNA" for the quark.

If quarks are to preserve their structure and withhold their energy, they must have an organizer of the strong nuclear force. In organized energy subsystems, like a proton or a neutron, the element also has a special local organizing quark, an alpha quark that works with the atomic quark. The atomic quark manages all subsystems, while the alpha quark organizes all quarks and functions of the local subsystem (proton/neutron).

If a local energy system (proton/neutron) loses its connection with the atomic quark, the alpha quark can keep the local system together for a short while. Eventually, due to its limited control over the strong nuclear force, it will fail to maintain cohesion, and the local system will dissolve. The local element will dissolve, and all the quarks will, in a short while, also dissolve and release their energy.

In order for a "free" quark to remain stable, it must rapidly connect with one or more other quarks to form a new coherent energy structure—such as a photon or other composite quark system.

Chapter 14

Particles

A particle in the Ilefos model share some of the properties of an atm, like mass, but does not have a complete atomic system. When two or more quarks connect and form an organized system, it is called a particle, until it has grown enough to develop a full atomic management system.

A particle has mass, is affected by gravity (ilefos tracks), and contains stored energy (dielectricity). However, it has weak mass and energy functions. It does not have enough energy to form ilefos or other tracks. If a particle is exposed to high levels of energy, it can create new quarks and grow in size.

If a particle interacts with an advanced energy system with ilefos tracks, like an atom, its quarks' energy might be absorbed before before it is repelled. Advanced organized energy systems with great energy-handling capacities, like dark matter, might absorb all of the particle's energy and quarks.

In the chapter "Organized energy systems – building of matter," the growth process of energy systems will be explained in more detail.

Chapter 15

Photons

In the Ilefos model, photons represent the smallest organized group of quarks. For an entity to possess energy and mass, it must be able to concentrate and store specific forms of energy.

A quark is the organization and concentration of a specific energy. Nuclear quarks can also be responsible for a special task. An atomic quark manages and organizes a quark group and contains the atomic DNA. This quark, together with a special quark, holds all the instructions and software for the quark group.

Photons have mass and energy. Their energy is evident in how they supply heat to matter and how solar cells can supply electricity through the photovoltaic effect. The energy of photons also gives energy to plants through photosynthesis. When light (photons) passes strong gravity, we see how light bends in gravitational lensing. This shows that photons have mass and are influenced by gravity.

Photons, therefore, must have one or more energy quarks and a quark that is affected by gravity.

For quarks to hold together in an organized system, we need a binding energy—like a glue— that can hold quarks together. In other words, we need a strong nuclear force quark. In this model, the energy providing the strong nuclear force is YT energy, and the responsible quark is a YT quark.

The YT energy has a strong attraction to ilefos (gravity). This attraction influences ilefos tracks from atoms and explains why ilefos (gravity) tracks can provide both strong atomic bindings and weaker bindings, like we see in gravity.

The simplest group of quarks must have a YT quark to keep the energy quark(s), the YT quark, and other quarks together. When the conditions are right and a quark group is formed, an atomic or "managing quark" is also formed. The smallest quark group consists of an energy quark, a YT quark, and an atomic quark. All these quarks also have a special support quark, which contains the "software" and instructions for the nuclear quark. A photon, therefore, consists of one or more energy quarks, a YT quark (strong nuclear force), and an atomic quark (organizer quark). Photons can differ in their energy levels. The number of energy quarks in the photon and their energy level determine the photon's energy level.

In the next chapter, we will discuss the nature of photons and how photons can be both light particles and waves.

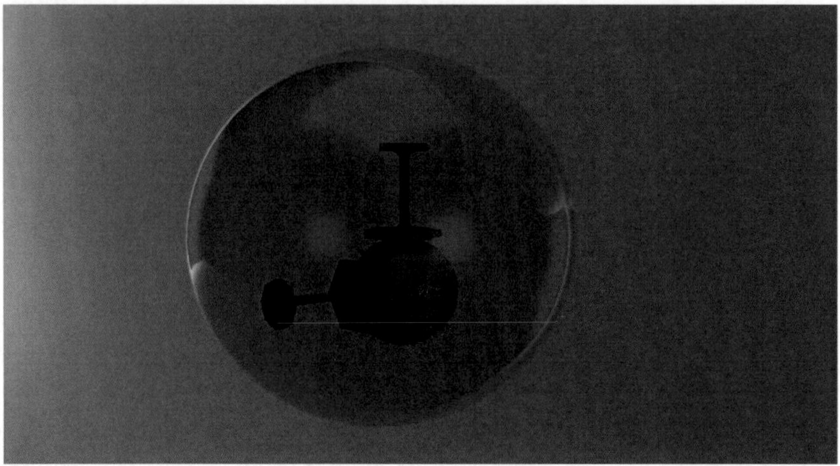

Figure 9. A photon consists of a YT quark (strong nuclear force), one or more energy quarks and an atomic quark. The particle may also have a few other essential quarks. (Credit: Stian Kongsvik).

Chapter 16

Light-Particles and Waves

In an article in Nature Communications (2 March 2015), Fabrizio Carbone and his team at EPFL Laboratory for ultrafast microscopy and electron scattering, captured a snapshot of light behaving both as a wave and a particle. This confirmed Albert Einstein's claim that light behaves as both a particle and a wave. Light is both a wave and a stream of particles.

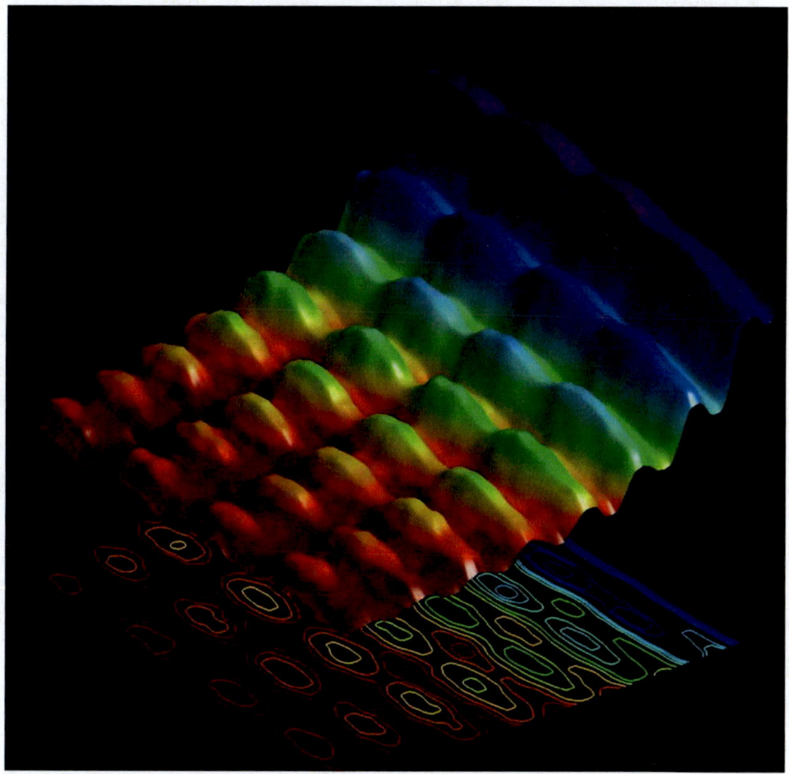

Figure 10. Photograph of light as both a particle and a wave, Fabrizio Carbone/EPFL.

But how can light be both a wave and a particle?

Waves − Frequency

If a photon consists of both energy quark(s) and a YT quark (strong nuclear force quark which attracts ilefos and external strong nuclear force), the photon is a particle with energy and mass properties. A photon is then the smallest group of quarks in an organized system. The system is too small to carry out complex functions such as energy conversions or forming ilefos (gravity) tracks. However, it does contain stored YT energy, and the photon is therefore affected by external ilefos attraction. This attraction is much weaker than larger quark groups and matter with ilefos tracks.

When photons are released from a source, several photons travel in almost the same direction and speed. The mutual attraction of the strong nuclear force (YT energy) causes photons traveling in the same direction, form small groups. The strength of the YT quark determines the strength of the mutual attraction between the photons and the size of the photon groups. The photon's energy level depends on the number of energy quarks and their energy level.

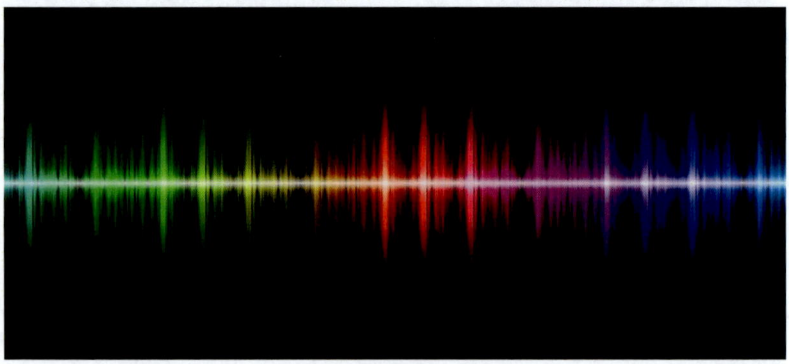

Figure 11. Waves, when photons travel in the same direction, their mutual attraction make them form groups of photons, which may be perceived as waves. (Credit: iStock, Hudiemm).

The frequency of light is defined by the number of photon passing per unit of time. The energy of a light wave is determined by the concentration of photons in the wave and their energy level.

We see how strong gravity (the attractive force as modeled by ilefos dynamics) bends light in gravitational lensing, which shows that photons are affected by gravity.

Figure 12. Gravitational lensing - strong gravity bends light. Credit: NASA James Webb telescope.

In light, groups of particles may be perceived as waves. Other particles and groups of matter that also have YT quarks (strong nuclear force), especially groups with plus (ilefos/gravity) quarks, also attract each other and form groups when they travel in the same direction. This attraction may form groups of matter, planets, and stars.

Waves can therefore be described as groups of particles or matter moving in the same direction due to mutual attraction. This can also be small repetitive movements in matter in permanent structures.

A concentration of free ilefos (gravity) energy units traveling in the same direction and are released in pulses may also be perceived as frequency. The ilefos energy units then arrive in altered strength, which are gravity waves. These gravity waves have been detected several times by LIGO and VIRGO detectors, the first time by LIGO in 2015. (Physical Review X, "Improved

Analysis of GW150914 Using a Fully Spin-Precessing Waveform Model," B. P. Abbott et al. (LIGO Scientific Collaboration and Virgo Collaboration), 041014 – Published 21 October 2016).

Waves can therefore be groups of particles that travel in the same direction and have mutual attraction to each other, forming groups or pulses of energy units. Such groups and pulses are perceived as waves.

Chapter 17

Universal Speeds

Speeds in the universe vary based on what moves, the environment, and the observation point. That which moves can be energy units or organized energies. Organized energies are energy concentrated and organized in systems, such as quarks, quark groups, particles, matter, and universal phenomena like stars and black holes. To measure speed, it must be observed relative to a specific point or location. We also have to determine the nature of the space itself. Is space fixed or a dynamic substance? Is space a dynamic fabric that expands, or not?

Universal speed depends on:

- The nature of what is moving
- The energy properties of the moving
- The energy properties of the environment, the nature of the fabric in which it moves in

General Relativity

First, in relation to space, we have to determine the dynamics of the "fabric of space." Is it a static fabric or a dynamic fabric that can expand and contract?

According to Albert Einstein, the speed of light is the universal speed limit. Nothing can move faster than the speed of light. To explain the expansion of the universe from the point of origin, the Big Bang, he has introduced the fabric of time and space, spacetime.

The universe expands at a speed much higher than the speed of light. This seemed inconsistent with the idea of the speed of light as an absolute limit. To explain the expansion, he introduced the fabric of space as a dynamic substance. In this explanation, he could also keep the speed of light as the universal speed limit. This substance, the fabric of time and space, expanded. He could now explain the expansion of the universe without coming into conflict with the speed of light as the absolute speed limit. Albert Einstein therefore introduced the concept of a dynamic element, the fabric of space.

This has been the general explanation of how the visible universe expands. Albert Einstein's general relativity gives good calculations of gravity and visible observations in the universe. But we have observations close to the center of galaxies where Isaac Newton's law of universal gravitation seems better to explain observations (Jody A. Geiger: "Discrete Approach to Star Velocity Resolves Dark Matter Phenomenon," ResearchGate, 2021).

The theory of general relativity has several weaknesses. The theory only works with astrophysics; it cannot explain the dynamics of nature's smallest elements, and it does not work with quantum dynamics. Correct physics should work in all fields.

The expanding universe is filled with galaxies and clusters of matter that don't expand. These structures then appear as mosaics of non-expanding space within an expanding universe. If the fabric of space expands, all space should expand. Also, what is the unknown force which pulls down and folds the fabric of spacetime? Is the universe flat? The theory of general relativity is based on a hypothetical fabric nobody can explain; it does not work consistently throughout the universe, and is not universal; it only works in astrophysics.

These points indicate that the theory of general relativity may be wrong. I will, therefore, work with a static, non-expanding fabric of space, and I will try to explain the dynamics of the universe through a different approach, based on energy and the dynamics of energies.

The Ilefos model is based on a universe in energy balance. A universe in energy balance means the energy in the universe is constant. In this model, "empty" space is filled with free energy units, dark energy. Energy can be organized and converted, but the level of energy is constant. General relativity is based on a hypothetical space with an evolving energy level, which is a universe not in energy balance. I prefer a balanced model, which I will explain in this book.

Speed of Organized Energies (Matter and Particles)

If an object moves and experiences resistance (drag), this will reduce its speed. The main source of resistance in the universe is the attractive force. The attractive force, ilefos, creates ilefos (gravity) tracks, which in turn make atomic bindings and gravity. The ilefos (gravity) tracks are strong close to the source; the attractive power is reduced with distance in the track. At the end, the ilefos track becomes too weak to hold together and dissolves into free ilefos energy units.

An object's drag of ilefos (gravity) depends on:

1) The strength of external ilefos force tracks.
2) The properties of the object moving – the object's attraction towards ilefos.
   - The strength of the object's ilefos tracks (if it has ilefos tracks).
   - The internal attraction of the ilefos energy and strong nuclear force (YT energy) in the quarks.
3) The concentration of free ilefos energy units (dark energy) in the environment.

An ilefos track is strong close to the source. The attraction, or concentration of ilefos energy units in the track, is reduced with distance from the source.

When two atoms' ilefos tracks connect close to the source, they can create a strong atomic binding like we see in metals. This creates a strong resistance towards movement. When atoms' ilefos tracks connect at a greater distance from the source, the tracks are weaker and form a weaker connection, gravity. Gravity exerts a weaker resistance- or drag-on a moving object.

"Empty" space is filled with dark energy. Dark energy is a free energy unit that is not connected to matter or universal phenomena. Free ilefos energy units contribute to a large part of dark energy. Individual free ilefos energy units also attract with an attractive force, but this force is very weak. In empty space with no ilefos (gravity) tracks, the attraction of free ilefos energy units constitutes all drag to moving elements. The concentration of dark energy can vary in the universe. Therefore, the drag of dark energy can vary, even though this force is very weak in our visible universe.

The moving object's attraction towards ilefos. There are three major energies that attract to ilefos:

- Ilefos toward ilefos
- The strong nuclear force toward ilefos
- Dielectricity toward ilefos

If the object is an organized quark collection, like we see in atoms, the object has ilefos tracks. Atoms, therefore, have ilefos tracks around themselves, which grip external ilefos tracks and attract free ilefos energy units. Matter and dark matter, therefore, have a strong drag from ilefos, which greatly reduces maximum speed.

Smaller quark groups that have not yet evolved to an atomic structure may have ilefos quarks (plus quarks). They do not have ilefos (gravity) tracks. They also have an attraction towards external ilefos and experience drag, which also reduces maximum speed. This drag is weaker than the drag of ordinary matter, and small quark groups (resa) can therefore move faster. (Resa is a small group of quarks, which are explained later in the book.)

Our smallest groups of quarks, photons, consist of energy quarks and a YT quark. The YT quark is responsible for the strong nuclear force, which holds the small quark group together. The strong nuclear force also attracts ilefos. This attraction is weaker than the attraction of larger quark groups with ilefos quarks (plus quarks), but photons are weakly affected by ilefos drag. We see this with strong gravity, which bends light in gravitational lensing. The drag/attraction toward ilefos in photons is very weak. Therefore, photons can travel much faster than larger quark groups, like atoms.

The speed of matter/group of quarks depends on:

- The properties of the quark group
- The environment in which it moves

Speed of Light

Light is the movement of photons. Photons, in the Ilefos model, are the smallest organized quark groups. They are the most basic quark groups, which consist of two types of nuclear quarks: Energy quark(s) and a YT quark.

The energy quark is responsible for storing and concentrating dielectricity. Dielectricity is the main energy for energy transfer between atoms, and it is the easiest energy to convert in matter.

The YT quark is responsible for the strong nuclear force. This force keeps the quarks together in a small, fixed structure.

In addition, all organized quark groups must have an atomic quark. The atomic quark is a management and control quark, which controls the actions of the group. The atomic quark also has the atomic “DNA,” which allows the quark group to evolve/grow if the environment allows this.

The strong nuclear force has an attraction toward ilefos. The YT quark (strong nuclear force quark), therefore, experiences a weak drag towards gravity tracks and dark energy.

We see this in gravitational lensing, where strong gravity from stars and black holes bends light. This also implies that photons are subject to a Doppler effect. Photons move a tiny bit faster towards strong gravity and a tiny bit slower away from strong gravity. This effect is small, but the nature of this effect tells us that the speed of light is not fixed; it can have slight variations.

The small drag of photons reduces the maximum speed of light in "empty" space. The speed of light in space is approximately 300,000 kilometers per second (299,792,458 meters per second), or 186,000 miles per second. The drag of the photons prevents them from reaching a higher speed. This speed can have slight variations based on whether they travel towards or away from strong gravity.

The strong nuclear force in photons also attracts the strong nuclear force of other photons. This attraction makes photons that travel in the same direction form small groups of photons. Groups of photons might be perceived as waves.

Different photons can have different quark compositions. This can give the photon a different level of the strong nuclear force, which in turn affects the size of groups of photons traveling in the same direction and the wave properties.

Speed of Energies

Matter experiences the drag of the attractive force, which limits the maximum speed. Matter consists of concentrated energy of different types in quarks.

Pure energy units, pulses of energy, which are not concentrated/organized, experience much less of the attractive force from gravity tracks and dark energy. The speed of energies is determined by:

- The properties of the energy
- The environment

First, energy units, no matter the type, travel faster when they are not concentrated/organized in quarks. Their attractive properties are much weaker when they are not part of an organized structure.

Second, different types of energies have different properties. Some energies have a strong attraction toward specific energies, while some have very weak, if any, attraction toward major attractive energies.

The dominating attractive energy is *ilefos.* Ilefos energy units attract each other and form links of ilefos energy units when released from the source.

Links of ilefos energy units are gravity tracks. Gravity tracks attract each other and may form connections like we see in atomic bonds and gravity.

Ilefos energy units have a strong attraction toward external ilefos. The ilefos energy units also have a strong attraction towards the strong nuclear force. This attraction reduces the speed when the ilefos energy units are released from a source that contains the strong nuclear force, like atoms. The influence of the strong nuclear force will be reduced with distance from the source. Ilefos energy units in a gravity track will therefore slowly accelerate with distance from the source. They will continue accelerating in the gravity track. This creates longer distances between the ilefos energy units, thus a weaker attraction between the units. In the end, the distance becomes too long, the mutual attraction of the units will not be strong enough to keep the units together on a track, and the track dissolves. The ilefos energy units become free ilefos energy units with no connection to the source. These free ilefos energy units are a large part of dark energy and are a major reason for the drag of dark energy in space.

The attraction of ilefos towards energies:

- Ilefos – ilefos, strong attraction
- Ilefos – the strong nuclear force, strong attraction
- Ilefos – dielectricity, moderate attraction
- Ilefos – most other material energies, low attraction
- Ilefos – dimensional energies, no attraction

I focus on the ilefos energy, which is the major influence on speed.

The strong nuclear force is not a part of dark energy, it only consists in organized energy systems, and it has a limited range around these. Therefore, this energy has almost no influence on speed in the universe.

Dielectricity energy units are also a large part of dark energy. This energy, therefore, has some influence on the speed of energies. They especially attract ilefos- and dielectricity energy units.

Other material energies have almost no influence of energy speeds. They have a low attraction towards other energies and are a small part of dark energy. Other material energies are energies concentrated in different quarks in matter and are a part of the atomic dynamic system.

Dimensional energies are energies that are not directly connected to material energies. These energies have no attraction toward material energies and therefore have no speed limit. Even though they are not directly connected to matter, they can have some touchpoints.

The estimated propagation speeds of different energy types are listed below. These values are theoretical and based on the interaction strength and degree of structural organization in each energy form. Here, "*c*" refers to the speed of light in vacuum (~299,792,458 m/s). Dimensional energies are modeled as non-local field synchronizations and are therefore not subject to conventional propagation limits.

Energy - speeds in space (estimates)

Energy	Estimated Relative Speed	
Ilefos	Moderate	6 c
Dielectricity	Moderate/high	11 c
Secondary material quark energies	High	15-99 c
Dimensional energies	Not subject to speed limits	Non-local synchronization; not materially bound
		c= speed of light

These speeds are theoretical estimates based on the interaction strength and structural organization of each energy form. "c" refers to the speed of light in vacuum. Dimensional energies are modeled as non-local field synchronizations and are not bound by conventional propagation limits.

The dielectric field response may propagate at effective rates estimated above the speed of light, e.g., ~11c, understood here as a non-material, coherence-based field modulation.

Ilefos pulses are modeled to travel at a coherent field propagation rate of ~6c, representing resonance pulses rather than classical particles.

Dimensional energies are modeled as non-local field synchronizations rather than traveling signals, and are therefore not subject to conventional speed limits.

These speeds are estimated speeds of free energy pulses. We still have no methods of measuring the speed of energies, and these are theoretical estimates derived from the modeled properties of each energy type.

Energies create forces based on their properties and interaction with one another.

The speed of the attractive force of ilefos (gravity) seems to be much faster than the speed of light as modeled in this framework. This can explain the accelerating expansion of the visible universe in relation to Big Bang ground zero. The visible universe now travels at a speed higher than the speed of light.

The dynamics of the accelerating expansion of the universe will be explained later in the book.

Gravitational Waves

Gravitational waves travel at a speed almost equal to the speed of light. We have seen gravitational waves detected by LIGO arrive almost at the same time as the photons from the merger of two black holes. This should indicate that gravity, ilefos, travels at the speed of light.

The nature of these gravitational waves was: They were the very strong waves of ilefos energy units. The waves were very large, dense groups of ilefos energy units. A large group of ilefos energy units that travel in the same direction experiences more drag from external energy units than single ilefos energy units. They clump together as energy clusters of ilefos energy units when moving in the same direction and speed. An energy cluster of ilefos energy has a larger drag due to the concentration of ilefos in the cluster. This causes large clusters of ilefos energy to units to move more slowly than individual units.

The nature of an ilefos energy unit is also that it sends out the attractive force. This attraction causes an ilefos energy unit to affect the area around it. This attraction allows ilefos energy units to form ilefos (gravity) tracks, which release concentrated areas with ilefos units when the tracks dissolve. These clusters of ilefos energy units travel as gravitational waves. Gravitational waves exert drag on other energies and organized energy systems.

It is this attraction, the attraction of ilefos, which travels fast at 6 c, not necessarily the units themselves.

This attraction works around the ilefos energy units with a limited reach. When a group of ilefos energy units travels as a gravitational wave, the field of attraction travels with the wave.

The observation of gravitational waves concerns waves (lumps) of ilefos, while the gravitational field itself moves independently of these lumps.

Chapter 18

General Relativity vs Universal Physics

The theory of general relativity was published by Albert Einstein in 1915 and is the universally accepted explanation of gravity in astrophysics.

General relativity provides accurate calculations of gravitational effects in the universe, but has weaknesses in explaining the underlying cause of gravity.

In general relativity, gravity results from the curvature of the fabric of spacetime. This model is often visualized as a flat (2D) surface that bends under the influence of mass, creating the illusion of a downward-pulling force.

The universe is not flat- and what exactly, is pulling "downwards" in a three-dimensional space? This theory, therefore, lacks a logical explanation.

A correct explanation must work in all directions and all three dimensions. If the object's properties pull in all directions, we have different objects that pull each other through gravity tracks (ilefos) from the various objects. Objects can be particles, matter, and universal phenomena (stars, black holes).

There is no clear explanation of how the 'fabric of space' functions physically, especially when modeled as a flat surface. The Earth is not flat, nor is the universe.

The expansion of the universe, an accelerating expansion faster than the speed of light, is explained by the expansion of space itself. The fabric of space, an unexplained fabric, expands. If space expands, why doesn't all space expand? In areas with matter, space does not expand. These regions then appear as mosaics of non-expanding space within an otherwiseexpanding universe.

General relativity gives us very good calculations of gravity in the universe. But it does not work in the quantum field, with nature's smallest objects, and it seems to lack a correct cause-and-effect explanation of gravity. A consistent physical theory must be compatible with both quantum mechanics and astrophysics. A correct physics with a clear cause-and-effect explanation could, with a good mathematical model, evolve to give a better description of gravity in the universe. In this book, I introduce the Ilefos model for a universal explanation of the attractive force in the universe.

Chapter 19

Organized Energy Systems – Building of Matter

Energy Growth

Different forms of energy can be concentrated and organized into energy systems. A specific energy can be concentrated in a quark. Quarks can form groups of organized energy, ranging from small, loosely bound systems such as photons, to larger, more complex structures like atoms.

In regions with high concentrations of a specific energy, quarks may form. To preserve their structure, quarks must form organized quark groups. An organized quark group has to have:

- An energy quark
- YT quark (strong nuclear force) to hold the group together
- An atomic quark that manages the organized energy system.

Organized energy systems can evolve when exposed to high-energy conditions. If an organized energy system receives more energy than it can store, it can form more quarks and grow. The quark group can develop more of the existing quarks, or the atomic quark can create new types of quarks. This energy-driven growth may continue until the quark group forms a complete proton. When the group has formed a proton, it has achieved all the functions of an atom.

Atoms can continue to grow through energy growth or fusion This process is further discussed in the chapter "The Growth of Atoms."

Chapter 20

Dissolvement and Energy Release from Matter

Atoms and other organized energy systems can both store and release energy. Albert Einstein showed us the mass-energy equivalence formula $E=mc^2$, which demonstrates the equivalence of mass and energy. We will take a look at how matter can release energy.

Friction

Atoms have attractive force which forms ilefos (gravity) tracks which pull atoms toward each other. To preserve their structure, the atoms must prevent collisions with other atoms. Collision between atoms can knock loose atomic elements (protons/neutrons) and may destroy the atomic structure.

To prevent this, atoms have a repulsive force. The repulsive force is the force that particles with repulsive force exert on each other in orbit around the nucleus. I call these particles uni particles and the repulsive force uni power.

A uni particle is a small quark group centered around one or more TE (uni) quarks. A uni particle has attraction towards the strong nuclear force, which holds it in a stationary orbit around the nucleus. It has a strong repulsive force toward other uni particles.

All atoms have a "shield" of uni particles which orbit the nucleus. They have a repulsive force toward uni particles in orbits around external atoms. This helps atoms keep a distance and thus prevents collisions between atoms.

Uni particles constantly interact with both the host atom and external uni particles. If atoms experience an increased force which pushes two atoms toward each other, the atoms must increase the repulsive force of the uni particles to prevent a collision.

The increased repulsive force helps keep a distance between the nuclei. This increases the uni power (repulsive energy) in the immediate area around the atoms, which also drains energy from the atoms. This increases the number of free uni energy units (TE energy) and pulses of uni energy in the area. Free energy units become a part of dark energy and can be absorbed by external atoms. External atoms then experience a strong increase in received uni energy

(TE energy). Atoms can transform uni energy/TE energy into dielectricity, which they release in dielectricity tracks. Increased dielectricity can be perceived as heat.

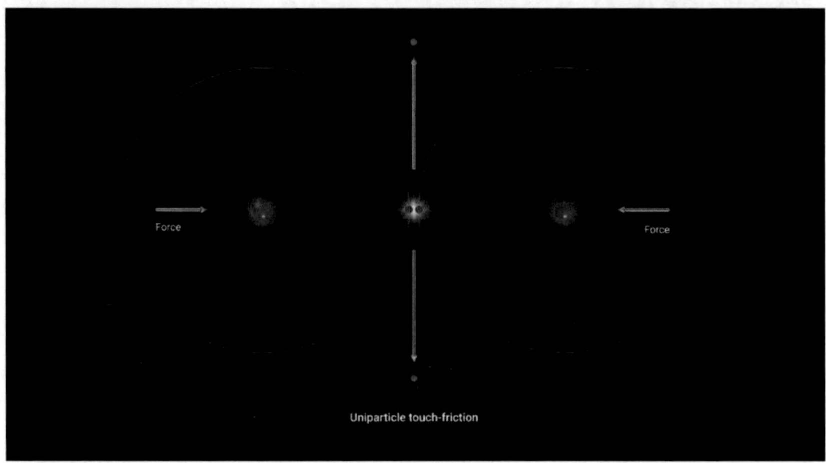

Figure 13. Two atoms increase the repulsive force to prevent collision. This dynamic interaction is what we refer to as friction. (Credit: Stig Boe/Mohammed Ismaiel).

Fission

When the repulsive force is not strong enough to prevent a collision between nuclei, we can have a strong or weak collision. A strong collision can result in atomic dissolution, which will be a topic for the next chapter.

A soft collision between two nuclei can result in:

- Some quarks in an atomic element might be knocked loose.
- An atomic element, like a neutron, might be knocked loose.

If some quarks are knocked loose from an atomic element, they will almost immediately dissolve and release their energy. Loose quarks without a management system and the strong nuclear force can't maintain their structure and dissolve. This results in an increase of free energy units in the surroundings of the atom. The resulting increase of dark energy (free energy units) may be absorbed by nearby atoms, providing them with additional energy.

The atomic element has enough energy to form new quarks and replace the missing ones. The atom continues as the same atom with the same number of atomic elements, but with slightly less energy.

If the soft collision is strong enough to knock a whole atomic element loose, like a neutron, this atomic element can maintain its structure for a while. An atomic element has an alpha quark, which runs the local operating system of the element. The element has all quarks intact, including a YT quark (strong nuclear force), and therefore can keep the structure as long as it has enough energy. The quarks have a lot of energy, but the loose element does not have the ability to harvest dark energy through harvesting tracks. In time, the element will therefore slowly lose energy and may dissolve and release its rest energy when its energy becomes too low.

A loose atomic element does not have uni particles, so other atoms' atomic shield (uni particles) will not repel a loose atomic element. A loose atomic element, like a neutron, therefore, might pass through the uni particle shield of another atom and collide with the nucleus. If this collision is strong enough, the collision can again knock loose an atomic element from the new atom. This released atomic element may in turn dislodge another atomic element upon collision with a different atom. We then have a fission process, a chain reaction like we see in a nuclear bomb.

An atomic bomb is, in the Ilefos model, a fission process of heavy energy-rich atoms that is triggered artificially.

Controlled fission, like we see in nuclear power plants, is actually a strong friction process among heavy elements, combined with some soft collisions that knock loose quarks. The heavy atomic elements are able to maintain their structure but will, in time, have reduced energy levels.

When an atom loses an atomic element, two things can happen:

- The atom can replace it and form a new atomic element.
- The atom can continue as an atom with fewer atomic elements. It can then turn into a new atomic element or become a new isotope of the same periodic element. An isotope is an element with the same number of protons, but with a different number of neutrons.

Strong, heavy, and energetic elements, like radioactive elements, have an abundance of energy and are able to replace their missing elements. "Weaker" elements might not be able to replace the missing atomic element.

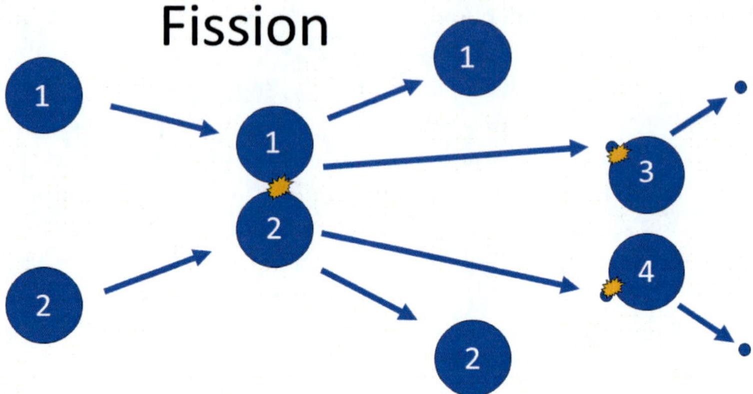

Figure 14. Fission. Two atoms, atom 1 and atom 2, collide, and an atomic element is released from each atom. The loose atomic element might hit new atoms, atoms 3 and 4, which in turn knock loose atomic elements in a chain reaction. (Credit: Bent Rolf Pettersen).

Atomic Dissolution

To grow matter requires energy. Albert Einstein's formula $E=mc^2$ (mass-energy equivalence) shows us the stored energy in matter.

When we dissolve matter, we can release this energy. In atom dissolution, the atomic elements (protons/neutrons) dissolve and release their quarks, which, within a short time, release their energy. All energy in matter is released as free energy units.

With nuclear fission, only part of the atom's energy is released. Most of the structure is intact, although the atom loses some energy. Strong nuclear fission in a fission bomb might create a limited atomic dissolution. In a hydrogen bomb, according to the Ilefos model, a nuclear fission bomb "ignites" a reaction with hydrogen isotopes, which then dissolve and release their energy. When the hydrogen isotopes dissolve and release their energy, this results in an energy release 100 times or more than that of the fission bomb.

When a loose atomic element from the fission process collides with a hydrogen isotope, the collision will most often destroy the hydrogen's proton structure. Within a short time, the proton will dissolve and release all its quarks, which in turn dissolve and release their energy. The hydrogen will not have enough energy to repair the proton. If the hydrogen atom also has neutron(s), it is an isotope. Then the neutron(s) also will dissolve and release their energy.

The atomic energy cycle is:

- Strong energy builds matter
- The dissolvement of matter releases energy

In stars, the main energy release is the dissolution of matter. The dissolution of atoms and subsequent energy release account for the immense energy output of stars. All matter and all quark groups that are absorbed by a star dissolve and release their energy. This includes dark matter.

The intensity, or power of a star, depends on the environment. How much and what kind of fuel is available gives the star its properties.

In a star, we have an atomic dissolution process that creates several hard collisions between organized energy systems (matter). The hard collisions crush the energy systems, which dissolve.

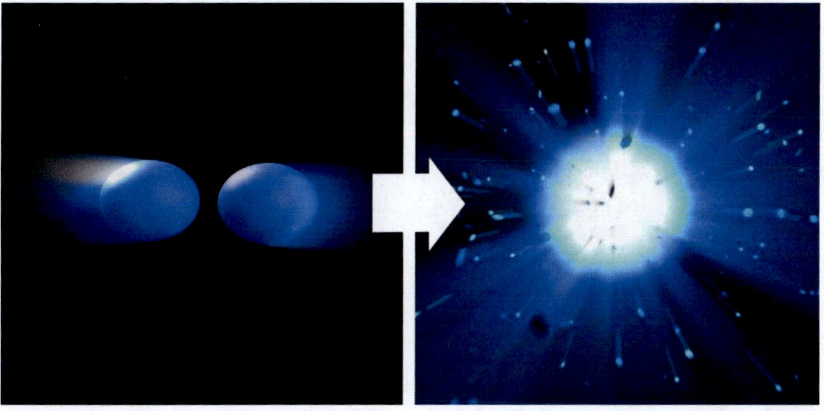

Figure 15. Hard collision between two atoms, which result in an atomic dissolution. (Credit: Ola Tandstad).

Inside a star, the strong energy gives an atom's uni particles (the repulsive shield) very wide orbits. The wide orbits allow the uni particles of different atoms to cross orbits, thus removing the protection of the repulsive uni particles.

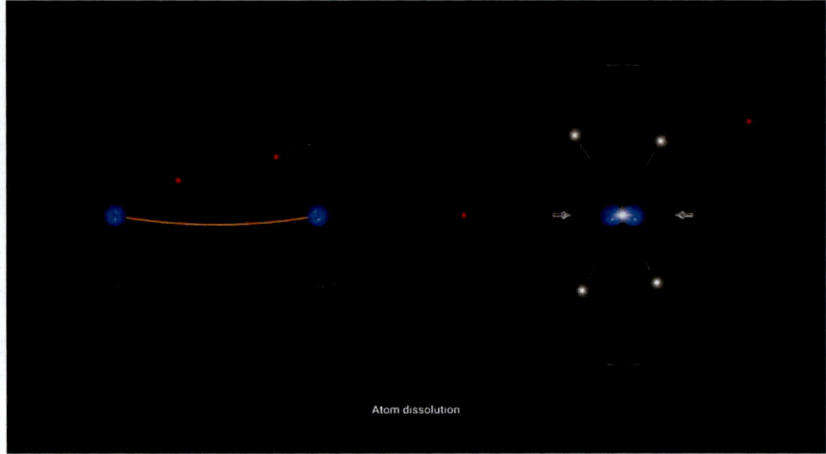

Figure 16. In an energy rich environment, uni particles get wide orbits. The uni particle orbits can then cross each other, and the nuclei can collide. (Credit: Stig Boe/Mohammed Ismaiel).

Fusion

Fusion is when two atoms merge into one larger atom. This can only happen in very energy-rich environments, in a soft nuclei touch.

Fusion requires the following conditions:

- Both atoms must receive large amounts of energy
- We must have a soft touch between the nuclei

Uni particles create a protective shield of repulsive force around the nuclei, which prevents them from touching each other. If an atom receives large amounts of energy, more than it can store, it must get rid of the excess energy. It does this in several ways. One way is to send more energy to the uni particles.

When the uni particles receive more energy, the more energetic uni particles travel in wider orbits. If the orbits widen sufficiently, the uni particle paths of two atoms may intersect. If this happens, the nuclei are no longer protected by a repulsive shield and might touch. If this is a very soft collision, the two nuclei might merge, and we have a fusion.

In the universe, this generally happens in very energetic environments when matter is traveling at almost the same speed and direction. The conditions are then right for a soft collision, which might result in a fusion.

In a fusion, two atoms merge and create a new atom. The new atom has all the atomic elements of the two atoms, unless one or both atoms have one or more extra neutrons. If the new atom has excess neutron(s), the extra neutron(s) will be released in the fusion process. When free neutrons are released in fusion, they will dissolve and release their energy within a short time. In this way, fusion can release energy.

- A fusion process between two atoms with no "extra" neutrons does not release energy.
- A fusion process between two atoms with "extra" neutrons releases energy.

In the chapter "The growth of atoms"/"Fusion," this process will be explained in more detail.

Explosion with Ordinary Matter

If an atom suddenly receives a large amount of energy, its quarks will be filled with energy. When the quarks reach their storage capacity, the atom has to get rid of the excess energy. The atom will then try to transport the excess energy away as dielectricity through its dielectricity tracks. The capacity of the dielectricity tracks is determined by the number of energy quarks (dielectricity quarks) in the atom. At the same time, some of the excess energy will be converted to uni energy (the repulsive force) and channeled to the atoms' uni particles.

The uni particles are particles in orbit around the nucleus with a repulsive force (uni power). Uni particles have a repulsive force toward external uni particles. This repulsive force acts as a repulsive shield that protects the atom from collisions with other atoms.

When the uni particles suddenly receive more energy, they will move faster in wider orbits with a stronger repulsive force. If this reaction happens to several atoms, all the atoms will have stronger uni particles, which travel in wider orbits. This will create a push. All the atoms will then try to push the neighboring atoms further away.

The atoms in metals and solid materials have strong atomic bonds that hold them together in a structure. Atomic bindings are strong ilefos tracks that connect close to the atoms where the ilefos tracks are strongest. The nature of an ilefos track is that the ilefos energy units are most dense close to the source. Then, the ilefos energy units will accelerate in speed, creating a longer distance between each unit. The ilefos track will become weaker with distance from the source.

If the increased repulsive force pushes the atoms further away from each other, the ilefos tracks, the attractive force, will become longer and weaker. If the repulsive force is strong enough, the longer distance between the atoms will create such weak atomic bonds that they break. The atoms then move away from each other without the atomic bonds. These "free" energetic atoms with no atomic bonds will, in turn, create a strong push on other atoms when they travel away.

The atoms may still have excess energy that must be released. The atoms create energy quarks, which they send out from the nucleus. These energetic quarks may form photons or other simple quark groups that travel away from the atom. These quarks and quark groups will create a strong energy transfer to external atoms. This might be perceived as heat and light.

This dynamic is what happens in an explosion.

What can start such a release of energy?

Oxygen is a material with great energy storage capacity. Oxygen has three energy quarks, which can store a large amount of energy. When oxygen quickly releases some of its stored dielectricity energy, it can result in the repulsive force and energy we see in combustion, gunpowder, and explosions.

For an oxygen atom to release some of its energy, we need a strong and rapid energy transfer to the oxygen atom. This "spark" provides the initiating energy required to trigger oxygen's energy release. When oxygen starts to release energy, we can have fire, combustion, and explosions.

When oxygen atoms receive this spark of energy, their energy quarks reach their threshold for storing dielectric energy. The result is a sudden release of larger amounts of dielectricity from the energy quarks. Some of this energy is transported away in strong dielectricity tracks. When the dielectricity tracks reach the maximum transport capacity, the atom uses the excess energy to produce energetic quarks, which are released. These quarks may form photons and other energetic quark groups, which may be absorbed by atoms nearby.

If this energy is sufficiently strong, it may trigger a repulsive reaction in surrounding atoms, as observed in explosions. The concentration of oxygen determines the energy level of this reaction.

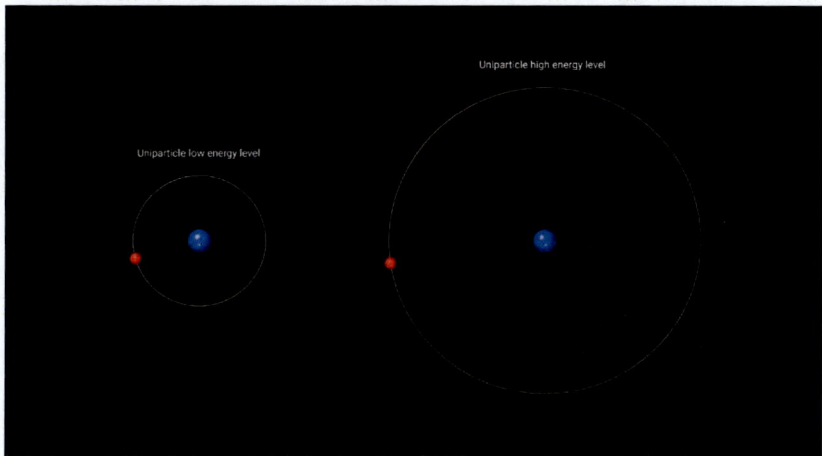

Figure 17. Atoms with low and high energy in uni particle. (Credit: Stig Boe/Mohammed Ismaiel).

Figure 18. When an atom receives large amount of energy, the uni particles also receive more energy. The more energetic uni particles then get a wider orbit. (Credit: Stig Boe/Mohammed Ismaiel).

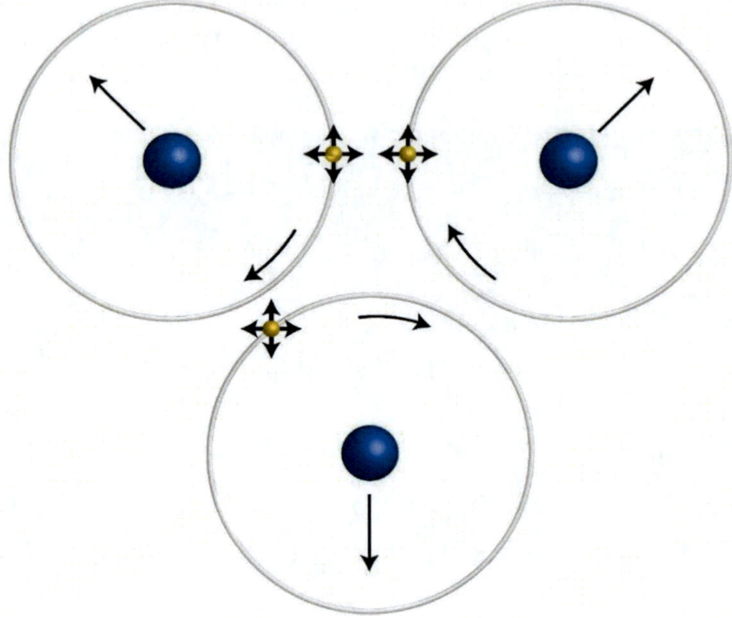

Figure 19. When atoms get more powerful uni particles in wider orbits, the repulsive force of the particles pushes the atoms further away from each other. This push can break atomic bindings if the ilefos (gravity) tracks become too long. (Credit Gary Cramer).

A similar process takes place in combustion and fire. Here, oxygen atoms release energy, which starts a similar process with hydrocarbons.

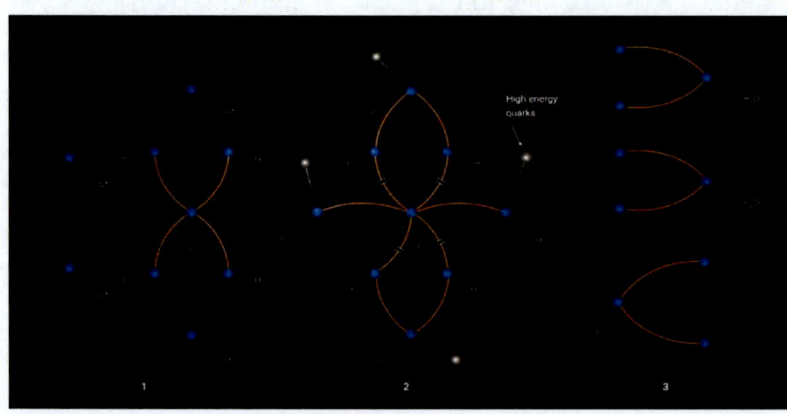

Figure 20. The combustion process between oxygen and methane. (Credit: Stig Boe/Mohammed Ismaiel).

In this example, a hydrocarbon (methane) and energetic oxygen atoms react. A "spark" makes the oxygen atoms' energy quarks reach their saturation point. They release energy in dielectricity tracks and as loose energetic quarks. At the same time, the oxygen atoms also get stronger ilefos (gravity) tracks, an attractive force, which makes them connect to the carbon atom's ilefos tracks.

Large amounts of energy are transported through ilefos and dielectricity tracks to the carbon atom. The increase in energy level in the carbon atom makes the carbon atom's uni particles stronger and move in wider orbits. This increase in repulsive force breaks the weak atomic bonds with the hydrogen atoms, which are repelled away. The free hydrogen atoms form atomic bonds with the now weaker oxygen atoms, forming H_2O. Two oxygen atoms form a new atomic bond to the carbon atom, C_2O.

Chapter 21

Preliminary Atoms − Resa

In the Ilefos model, photons are the smallest groups of organized quarks. A photon is a group of energy quark(s), a YT quark (strong nuclear force), and an atomic quark. This configuration has the most basic "atomic" operating system. It lacks the advanced properties of atoms but contains stored energy and mass, and it is affected by gravity due to the YT quark (strong nuclear force).

This basic quark group can grow to larger, more advanced quark groups. When a star releases a large amount of energy, the energy is very concentrated and forms different types of quarks. These quarks organize into basic quark groups, such as photons. If the concentration of energy and photons is dense, photons can merge into larger quark groups.

These larger quark groups possess greater mass, more energy, and are more affected by gravity. Because of their increased mass and attraction to gravity, they have a much slower speed than light. We can observe these particles entering the Earth's atmosphere as aurora borealis (northern light).

These larger quark groups are resa. Resa is the preliminary stage of atoms. Resa are quark groups that have formed a plus quark (ilefos/gravity). The ilefos quark increases the mass of the quark group. Resa can also grow to include most of the quarks of an atom.

We can divide resa into three types:

- Resa 0
- Resa
- Resa +

Resa 0

Resa 0 is the most basic form of resa. It has between 6 and 50 quarks. Resa 0 is loosely bound by a weak YT quark, and it has an ilefos quark, which gives extra mass. Resa 0 does not have ilefos tracks or uni particles (the repulsive force), and lacks many of the nuclear particles of an atom.

If a resa 0 is exposed to a large amount of energy, it can form new quarks. It does not have ilefos- or dielectricity tracks to eliminate excess energy; it must form new quarks to handle the energy. Its existing quarks have reached their energy storage limit and cannot store more energy. In this way, it can grow. It can also fuse with other resa if they have a soft touch.

Resa

Resa is a larger quark group with between 51 and 160 quarks. Here, we see the beginning of a proton, but it still lacks many of the nuclear quarks of an atom. It does not have an Ilefos- and dielectricity track, but it has formed a weak plus quark (ilefos quark). The YT quark is stronger, and the quark group is held together. The quark group also has a more advanced operating system and has a larger mass.

Resa may grow when exposed to high energy levels and can fuse with other quark groups.

Resa +

Resa + represents a nearly fully evolved atomic structure. It has between 161 and 255 quarks and an advanced operating system. This quark group has almost all the nuclear quarks of an atom and has many atomic functions. It has created a TE quark (uni energy). Resa, therefore, has a weak uni particle that creates a repulsive shield that protects the quark group. It has a strong YT quark (strong nuclear force) which holds the quarks together. Resa + has a plus quark (ilefos quark) but still does not have gravity and dielectricity tracks. It still lacks the final quarks required to form a complete proton and atomic operating system.

If resa + are exposed to large energies, they can form the last nuclear quarks and become a complete proton, thus becoming an atom.

Chapter 22

Atoms

Atoms are the first organized quark group with a complete management system. When a quark group has 256 or more quarks, the atomic quark (management quark) has all the quarks it needs to form a complete atomic element, like a proton. Hydrogen is the smallest element and consists of a single atomic unit—a proton.

An atom has fully atomic properties like:

- Ilefos (gravity) tracks
- Dielectricity tracks
- Uni particles (repulsive force/shield)
- Atomic Phase displacement
- Fully developed communication abilities
- An advanced operating system

(Hydrogen only has one ilefos- and dielectricity track and one uni particle).

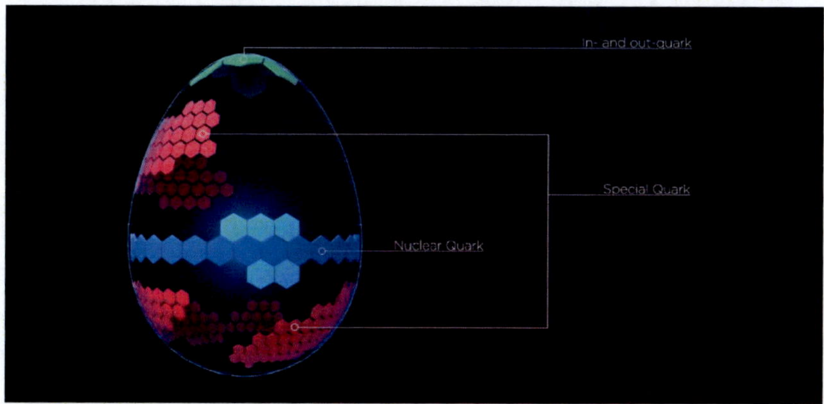

Figure 21. An atomic element (proton/neutron) is different quarks organized in a system by the atomic operating system and the strong nuclear force. (Credit: Ola Tandstad).

The strong YT quark provides the strong nuclear force, which holds the quarks together in an atomic element (proton/neutron). Atoms can have many atomic elements. Most of the time, the number of protons and neutrons is equal. The number of protons characterizes the matter in the periodic table. However, the atom can have extra neutrons and thus become an isotope.

Protons and neutrons have the same quark composition, but they behave differently. They have a local management quark, an alpha quark, which administers the atomic element. The atomic quark manages the atomic elements through their alpha quarks.

Protons

In the Ilefos model, protons prioritize dielectricity, communications, and dimensional energies. Protons emit dielectricity tracks. Each proton emits a dielectricity track if the atom has surplus energy.

The dielectric energy capacity of a proton depends on how many energy quarks (dielectricity quarks) it contains. An energy quark may correspond to what is known as the gluon in quantum chromodynamics. Protons can have from one to four energy quarks (gluons). The number of energy quarks in protons determines the thermal and electrical conductivity of the atom. Copper (Cu) has several energy quarks and is a conductor. Atoms with only one energy quark have low electrical conductivity and are insulators.

Hydrogen does not have a neutron. In this case, the proton also must handle the tasks of the neutron. This gives the hydrogen atom very weak atomic actions, but it is still a fully operational atom that can convert energies and perform all the other atomic actions.

Neutrons

Neutrons prioritize ilefos and TE energy (uni power). Each neutron sends out an ilefos (gravity) track and handles a uni particle (repulsive force). The power of the Ilefos track determines the strength and gravity of atomic bindings. A neutron with many plus (ilefos) quarks, can make strong atomic bindings with neighboring atoms. A plus quark may correspond conceptually tothe Higgs boson. Iron (Fe) has four plus quarks, prioritizes ilefos, and can have strong atomic bindings. Gases have only one plus quark and have very weak ilefos tracks. Helium (He) is an example of an atom with one plus quark.

Atomic Phase Displacement

Atoms must constantly convert energy to perform their tasks. Atoms constantly need a strong nuclear force for quarks and atomic structures to stay together as an organized energy system. The repulsive force of the uni particles must also be continually sustained and adjusted according to the environment. To communicate, atoms must also provide different communication energies, and the operating system (atomic quark) also needs a special energy. The attractive force (ilefos tracks) which perform atomic bindings and gravity, also demand a constant feed of ilefos energy.

The quarks responsible for the dedicated energy do not always have enough energy to perform their tasks optimally.

In order to perform and adjust the atomic tasks, the atom must constantly transform energies. The neutral quark is responsible for converting energies in an atomic element.

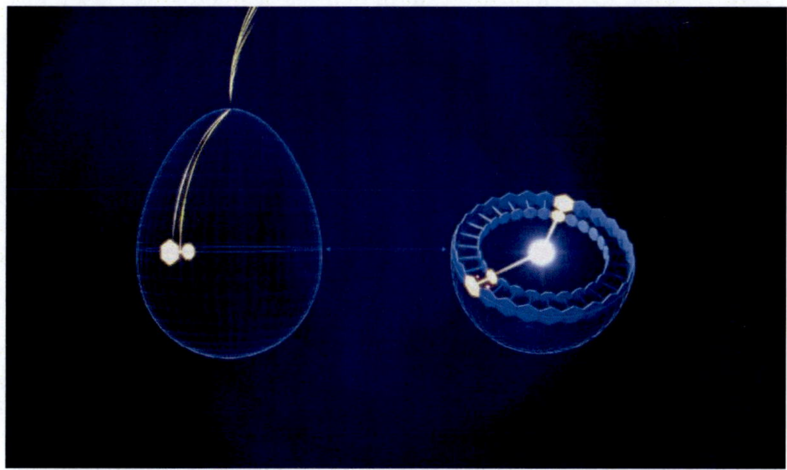

Figure 22. Atomic Phase Displacement in an atomic element. The neutral quark receives energy from a quark and converts it to a different form of energy. The converted energy is sent to a specific quark, which administers the energy. (Credit: Ola Tandstad).

Different atoms prioritize different energies. If an atom has several plus (ilefos) quarks, it prioritizes ilefos and ilefos tracks. If an atom has several energy (dielectricity) quarks, it prioritizes dielectricity and dielectricity tracks. When an atom receives more energy than it can handle, it must release the excess energy.

Iron has four plus (ilefos) quarks. If iron receives large amounts of energy, it converts the excess energy to strong ilefos tracks. Magnetism is very strong ilefos tracks (attractive force). In an electromagnet, iron converts electricity to ilefos (magnetism). Only other atoms with four plus quarks can fully utilize the binding strength of the strong ilefos track, as seen in magnetic materials. Copper has several energy (dielectricity/gluon) quarks. If copper receives large amounts of energy, for example ilefos from magnetism, it converts it through Atomic Phase Displacement to dielectricity, which it releases in strong dielectricity tracks. We see this in a generator.

In order to perform Atomic Phase Displacement, the atom has to have a specific quark that can interact with all energies. In the Ilefos model, this is the Neutral quark. The Neutral quark has a small amount of UR energy, which can interact with all energies. (Read more about UR energy in a separate chapter later in the book and in the chapter around dark matter.) Since UR energy can interact and attract all other energies, the Neutral quark can, with a specific operating system, convert all energies through the UR energy in the quark.

Energy Balance in Atoms

Atoms constantly use a lot of energy to perform their tasks. Atomic tasks are performing the strong nuclear force which holds the nucleus together, the attractive force (ilefos tracks), the repulsive force (uni particles) and several other internal tasks.

Iron, for example, uses a lot of energy to create atomic bonds with neighboring atoms and create gravity (the attractive force as modeled by ilefos dynamics).

To keep this up, atoms must be in energy balance. Atoms must receive as much energy as they emit.

Where can atoms receive energy? Dark energy. According to the Lambda-CDM model, 68 percent of the observable universe is dark energy. In this book, dark energy is defined as free energy units that are not connected to organized energy systems, like atoms. We must assume dark energy consists of the same energies as we find in matter, plus some dimensional energies.

Most of the environment of atoms is then filled with free energy units (dark energy). We therefore must assume that atoms constantly harvest free energy units in order to be in energy balance.

Atoms then have to have a harvesting track that harvests dark energy. Ilefos seem to attract some of the major energies, like dielectricity. These energies also constitute a large part of dark energy.

All atoms have at least one ilefos (gravity) track. Each neutron manages an ILEFO's track. An Ilefos track consists of Ilefos energy units released from a neutron, where the mutual attraction makes the units form a track. We can assume the attraction also attracts free ilefos, dielectricity and other energy units which are attracted by ilefos. If these units are transferred back to the atom, the atom has a constant feed of energy from free energy units in the environment. In this way, atoms can be in energy balance.

Atoms receive energy:

- From dark energy through harvesting tracks (Ilefos tracks)
- From dark energy directly to the nucleus
- From external dielectricity tracks
- From photons or other small quark groups

Atoms share energy through their dielectricity tracks, and atoms can receive free energy units directly to the nucleus.

Atoms generate both ilefos and dielectricity tracks. The energy units that travel in these tracks become free energy units when the tracks dissolve at the end of the track. Atoms, therefore, feed the universe with dark energy. This is also the case with gravity and other energy tracks from universal phenomena, like stars and black holes. Energy units are therefore recycled in atoms.

The nature of an ilefos track is that it is most dense with ilefos energy units close to the source, then becomes weaker the greater the distance between the ilefos energy units and the source. This means dark energy units, for example, free ilefos or dielectric energy units, become attracted to the attractive force of the ilefos energy units in the ilefos track.

The free energy units (dark energy) migrate toward areas of stronger attraction, moving along the gradient outside the ilefos track. The ilefos track is denser and has a stronger attraction closer to the source. The free energy units therefore move towards the stronger attraction and denser part of the ilefos track, which leads them to the source of the ilefos track, the nucleus, the neutron, which sends out the ilefos track. The steady flow of free energy units (dark energy) is then absorbed by the neutron, which gives the atom a stable supply of energy. When the neutron receives these energy units, a special quark puts them into the atomic energy system, where it is shared with protons and another part of the nucleus.

In this way, an atom can be in energy balance even though it constantly sends out energy.

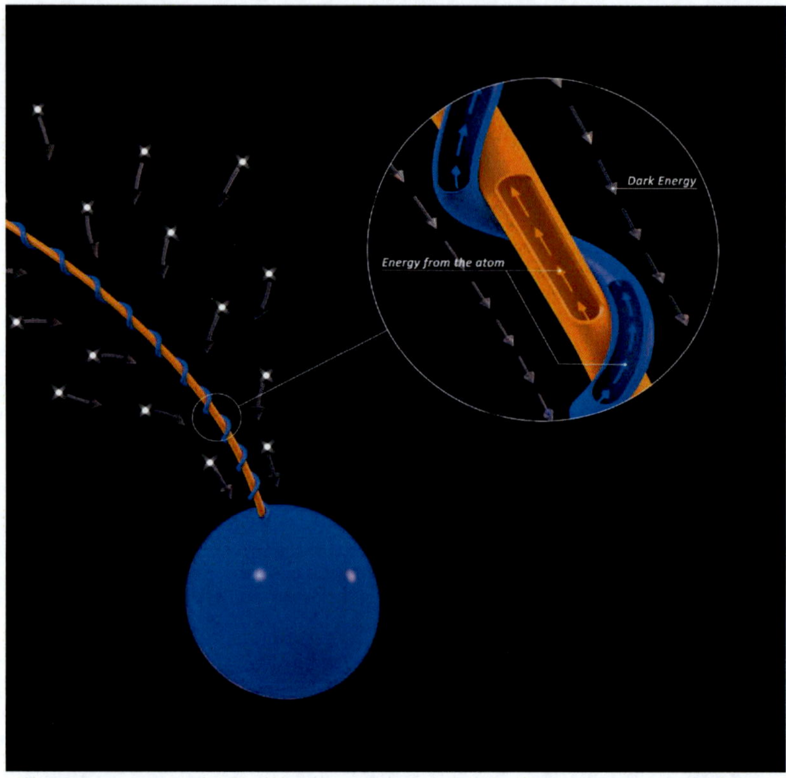

Figure 23. Energy balance in an atom. Dark energy units are harvested through the ilefos track to the nucleus. (Credit Ola Tandstad).

Chapter 23

Properties of Atoms

Atoms perform several functions to maintain their structure and enable key atomic processes:

- Produce the strong nuclear force to hold atomic elements together in a structure
- Produce the attractive force (ilefos), which creates atomic bonds and gravity
- Produce the repulsive force (uni particles) to protect the structure
- Atoms constantly convert energies through Atomic Phase Displacement
- Atoms communicate with each other and are aware of their surroundings
- Manage an internal operating system
- Atoms can create new quarks and atomic elements

Atoms have mass and energy.

Mass reflects how atoms are influenced by gravity or the attractive force. This arises from the presence of YT quarks, which generate the strong nuclear force, and plus quarks, which produce ilefos, the attractive force. Both of these energies are attracted to ilefos tracks (gravity tracks).

Energy is how atomic quarks can concentrate and store large amounts of energy. The most significant are energy quarks, which store and generate dielectricity. This can be emitted in dielectricity tracks or as energy quarks. However, all quarks contain energy that can be converted and released during fission or atom dissolution.

Most of the atomic actions have been explained previously in this book, but we have not yet described the communication of atoms.

Atoms continuously communicate, remain aware of their surroundings, and respond to environmental feedback. If atoms are moving towards each other, they must react to prevent a collision between the nuclei. An atom does this by increasing the repulsive force of the uni particles. An increased repulsive force helps the atoms keep distance and thus prevent a collision that could destroy their atomic structure. The increase in dark energy in the area, caused by uni particles, leads to friction.

In quantum entanglement, two particles are synchronized. They will then have a synchronized spin. The particles will continue to communicate when they are separated. If we change the spin of one particle, the other particle's spin will also change. We can see this effect over several hundred kilometers, and the speed at which they communicate is much faster than the speed of light. This suggests they communicate through a dimensional energy —an energy not influenced by gravity and capable of propagating at extreme speed. It is probable that atoms use different energies to communicate locally and over larger distances. Atoms may possess a specific dimensional quark for each of these communication pathways, enabling them to interact both locally and at a distance.

Chapter 24

Atomic DNA

In the Ilefos model, atoms possess an atomic quark that runs the atomic operating system. This quark communicates with all atomic elements and executes actions in response to feedback.

The atomic quark must also have a special quark that contains the operating system's software. This special quark contains both the operating program and the source code of the atom. The special quark must have the blueprint of the organized quark group, which also holds the blueprint for how the quark group may evolve. From photon to resa, atom, and even dark matter, all developmental instructions must be encoded in this special quark, which follows the quark group when it evolves. It carries the atomic DNA.

DNA means deoxyribonucleic acid, which carries the genetic instructions of living organisms. DNA instructs the organism how to develop, perform its functions, and grow. Atoms appear to exhibit similar functional characteristics. I refer to this as atomic DNA, although it is not an acid in the biochemical sense.

Chapter 25

The Growth of Atoms

Organized energy systems can evolve. Atoms may grow in two primary ways:

- Energy growth
- Fission

Energy Growth

By far the most common way for atoms to grow is through energy growth. If an atom receives large amounts of energy, it will initially store the energy in its quarks if this is possible. If the atom's quarks are fully saturated, it must release the excess energy. First, it tries to export energy through dielectricity and Ilefos tracks. The limit for energy in these tracks depends on the quark composition of the atom. At the same time, the atom also increases the energy in its uni particles- the repulsive force.

If this is not enough to expel the excess energy, the atom starts to produce and release energetic quarks, generally energy quarks. These energetic quarks may, in turn, form photons.

If the production speed of quarks is too high, the atom will be unable to expel them from the atomic elements. Each atomic element can only expel a limited number of quarks. When it exceeds this limit, the quarks must be stored internally in the atom. The atom then forms new atomic elements in the form of new neutrons and protons. It then grows in size and becomes a new atomic element or an isotope.

The atomic quark regulates growth based on the atomic DNA. The new atomic elements could take on a new quark composition and change the quark composition in all atomic elements.

Energy growth occurs in very energetic environments. This could be in exploding stars, such as a supernova, and in strong energy beams from quasars/Hawking radiation, pulsars, and other universal phenomena.

Almost all matter is created by energy growth.

Fusion

If the conditions are right, atoms might merge. For atoms to fusion, the following conditions must be in place:

- A very energetic environment
- They must undergo a very soft collision, an atomic touch

The atoms must receive massive amounts of energy, but not enough to start energy growth. The atoms then send much of the excess energy to their uni particles. The energetic uni particles will develop new and extensive paths, further repelling the repulsive uni particles from the nucleus. The paths of uni particles of different atoms could then cross, thus removing the atoms' nuclei protection. This allows the nuclei of the atomic nuclei to collide or make contact.

If the outcome is a brutal collision between the nuclei, we will see a fission process or atomic dissolution. If the result is a soft touch between the nuclei, they may merge in an atomic fusion.

To have a soft nuclei touch, the nuclei must preferably travel in the same direction and at the same speed. If the nuclei are not protected by the uni particles (the uni particle orbits have crossed each other) and travel in the same direction and speed, the outcome could be a nuclei touch and a fusion of two atoms. The two atoms then fuse into one larger atom, with all the atomic elements of the two former atoms. One of the atomic quarks (managing quarks) dissolves, and one atomic quark administers the function of the new atom according to the atomic DNA.

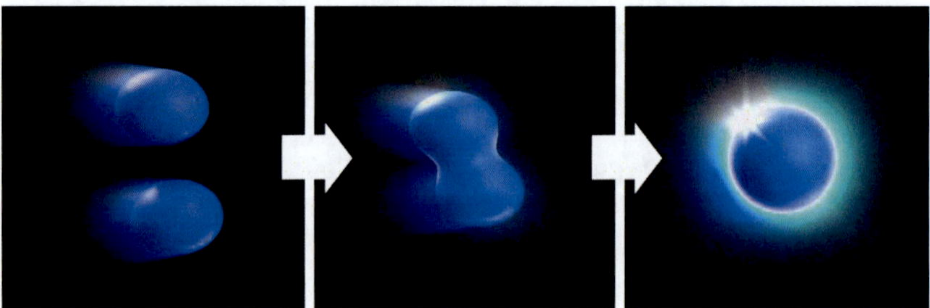

Figure 24. Two nuclei have a soft touch and fuse into a new atom. (Credit Ola Tandstad).

In a general fusion between two atoms, we have no release of energy. All atomic elements, quarks, and energy are retained in the new atom. Only the energy from the redundant atomic quark, which is dissolved, is released. Most of this energy is obtained by the new atom's quarks.

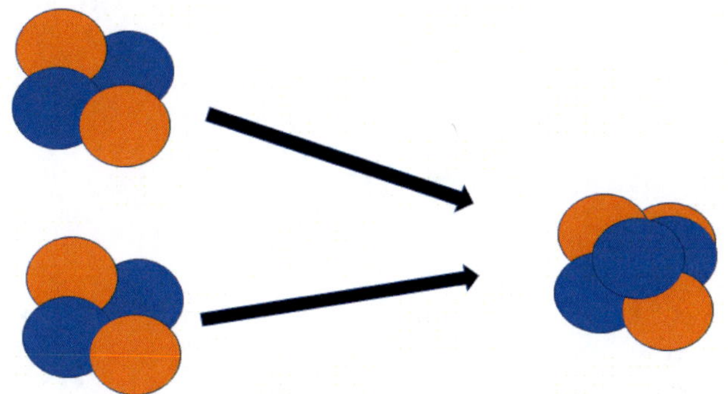

Figure 25. Two atoms with no extra neutrons fuse. (Credit: Bent Rolf Pettersen).

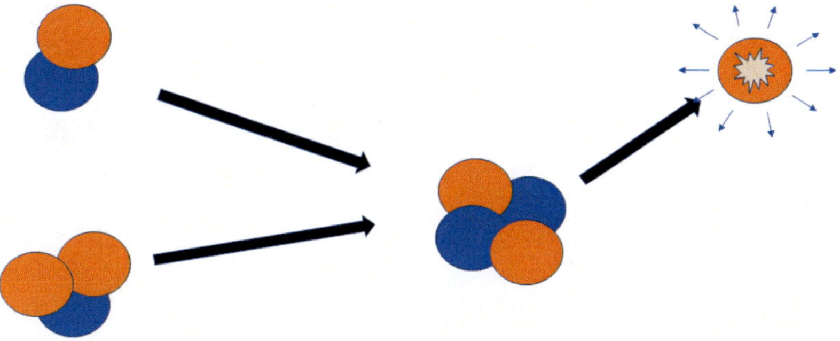

Figure 26. Fusion between two atoms. One of the atoms has an excess neutron, which is released, dissolves, and releases its energy. (Credit: Bent Rolf Pettersen).

If one or both atoms are an isotope with extra neutrons, the fusion process will release energy. When the two nuclei merge, they will have one or more excess neutrons. The excess neutron will be released from the new atom as a free neutron. The free neutron may maintain its structure briefly, but without an atomic quark (management), the strong nuclear force cannot hold the quarks together in a structure. The neutron will then dissolve and release the energy in its quarks.

In general:

- Fusion between ordinary atoms does not release energy
- Fusion involving atoms with extra neutron(s) (isotopes) releases energy

Chapter 26

Dark Matter – Dark Atoms

In the universe, we observe gravitational effects that cannot be explained through general relativity or other current theories. This material does not appear to interact with light or electromagnetic fields. This unknown material is referred to as dark matter.

In the Lambda-CDM model, which estimates the mass-energy content of the universe, 26.8% of the visible universe is dark matter, 68.2% is dark energy, and 5% is ordinary matter. Dark matter comprises 85% of the mass in the universe, and it is a substance nobody can explain.

Dark matter is believed to influence the creation of stars and galaxies and the movement of galaxies and stars in the universe. We can see it form a strong gravitational force through gravitational lensing. It is considered cold, meaning it emits neither heat nor radiation, only gravity (the attractive force as modeled by ilefos dynamics).

The properties of dark matter are:

- It has very strong gravity
- It does not reflect or emit light
- It does not emit heat
- We cannot visibly observe the material

If dark matter is a type of matter, we can assume it is composed like ordinary matter. But why does it seem to have different properties?

When matter grows, either through energy growth or fusion, the new expanded matter has new properties. If an atom is exposed to large amounts of energy, more than it can rid itself of through normal channels, it forms new quarks and new atomic elements (proton(s)/neutron(s)) and becomes a new, larger atom. The resulting atom may exhibit a new quark composition, giving rise to altered properties.

Is there a limit to the size of organized energy systems like atoms?

All growth of atoms occurs according to the atomic DNA, the blueprint of matter. If heavy atoms are exposed to strong energy, do they still grow? Can heavy atoms fuse?

It is likely that atoms can grow beyond the limits of the known periodic table, and then pass a threshold, which gives the new atom new properties. The new properties are the properties of dark matter.

If the atom grows beyond this threshold, we can assume that the atomic elements grow new types of quarks, enhancing some of the atomic properties.

The new dark matter atom has very strong gravity. It has much more powerful ilefos (gravity) tracks than normal atoms. These ilefos tracks extend significantly farther than those of ordinary matter. Therefore, the new atomic elements must have more plus quarks (ilefos quarks). It will then, as is the case with iron, prefer to exert an attractive force with an abundance of energy. This suggests that dark matter contains a higher ratio of plus quarks relative to energy quarks.

The long and strong ilefos tracks of dark matter also allow it to harvest large amounts of energy (dark energy), much more than ordinary matter. To be in energy balance, dark matter emits very powerful ilefos (gravity) tracks.

The material must also have a much higher energy capacity. It can absorb all energy from photons and does not therefore reflect light. This could explain why we are unable to observe dark matter.

But why do we not see the shadow of dark matter? If dark matter has strong gravity tracks, it should also form highly dense matter, the shadow of which we should be able to see.

If dark matter is enhanced matter, dark matter atoms may also have more and stronger TE quarks, which generate a repulsive force. Dark matter would, as such, have much stronger uni particles that travel in much wider orbits than normal matter. This would create a much longer distance between the dark matter nuclei, resulting in a material of extremely low density. Photons may thus pass through it if they do not encounter dark matter nuclei or the dark matter's ilefos tracks.

Dark matter is cold because it prioritizes ilefos tracks (attractive force) before dielectricity tracks. Its enhanced energy handling capacities also enable it to absorb and store all energy, except what it sends out as ilefos (gravity). It does not produce quarks if exposed to strong energy.

In the Ilefos model, normal atoms have between 256 and 290 quarks, divided between 23 types of nuclear quarks. When an atom passes 290 quarks, it passes the threshold and creates new types of quarks, creating dark matter. Dark matter has much stronger atomic elements (protons/neutrons) with more quarks in each element, but it also contains some new types of quarks and more atomic elements.

We must assume that dark matter atoms can also grow if exposed to strong energy, such as an energy beam, but only within a specific limit. If the dark matter atoms become very large, they can rapidly absorb all the energy in the energy beam without creating new atomic elements.

Why is it not the case that normal matter sticks to dark matter due to its strong gravitational force, thereby rendering it visible?

In a binding between two ilefos tracks, the strength of the binding is determined by the weakest ilefos track. We also see this in normal matter, where magnetism (strong ilefos tracks) can only form strong bindings with ilefos tracks of material with the same ilefos capacity (magnetic materials). Only iron-like atoms, with four plus quarks (ilefos quarks), can produce magnetism, and only the same types of atoms have magnetic properties, enabling them to form strong connections. This is also the case with the strong ilefos tracks of dark matter. These strong ilefos tracks can only form strong bindings with other strong ilefos tracks from external dark matter and universal phenomena (stars/black holes).

The Special New Dark Matter Quark

A dark matter atom has immense energy capacity. It can store much more energy than normal atoms and thus absorb all the energy of photons and other quark groups without uni particles. Dark matter has very strong uni particles (repulsive force) moving in very wide orbits, hindering a merge between normal atoms and dark matter.

If dark matter only had more energy (dielectricity), quarks, it would not be able to explain its energy storage capacity. To explain the nature of black holes, this model introduces a distinct energy that can affect all other energies, known as UR energy.

UR energy is the primordial energy. UR energy can attract all energies and therefore manage energies better than YT energy (the strong nuclear force). UR energy can easily be converted into all other energies. In normal atoms, the neutral quark manages Atomic Phase Displacement, or conversion of energies. We must assume that a neutral quark has a small amount of UR energy, allowing normal atoms to convert energies.

If dark matter atoms have separate UR energy quarks, they could store this energy with the properties of all other energies. UR quarks can store all forms of energy, which are automatically absorbed and transformed within them.

Dark matter also has all the other quarks of normal atoms, but the UR quarks enable strong energy storage and handling capacities. They also enable strong Atomic Phase Displacement capacities, enabling immense energy conversion capabilities. This allows the dark matter atoms to produce very strong ilefos energy, resulting in powerful and long ilefos (gravity) tracks. It also provides the conditions for powerful uni particles (repulsive force), which helps the dark matter atoms to maintain a great distance between their nuclei.

When atoms grow to a certain size, the atomic DNA automatically creates UR quarks, creating dark matter atoms.

How to Grow Dark Matter

Atoms can grow in two ways:

- Growth by energy
- Fusion

When the heaviest atomic elements receive very strong energies that they are unable to rid themselves of through normal channels (dielectricity and ilefos tracks), they start to produce energetic quarks, which they then expel. If too many quarks are produced, they can form new atomic elements (protons/neutrons), which allow them to grow in size.

If the atomic element reaches a threshold, the strong nuclear force (YT energy) will not be strong enough to keep all elements together in a structure. The atomic quark, the operating system of the atom, will then, through the atomic DNA, start to produce the UR quark to maintain the atomic structure. The UR energy of the UR quark(s) has much stronger energy control capacities than the strong nuclear force. UR energy then starts to organize the energies, quarks, and atomic elements. This allows for larger atomic structures with more than 150 protons. The formation of dark matter is suggested to begin at a structural threshold of approximately 150 protons.

When two heavy atoms travel in the same direction and receive large amounts of energy, their uni particles will travel in very wide orbits. If these orbits intersect, we can have a soft collision that might result in a fusion between the nuclei.

If this fusion results in an atomic element with more than 150 protons, the strong nuclear force (YT energy) will not be able to keep the atomic structure together. The atomic quark then initiates the formation of a UR quark to maintain the structure of the new atomic element, a dark matter atom.

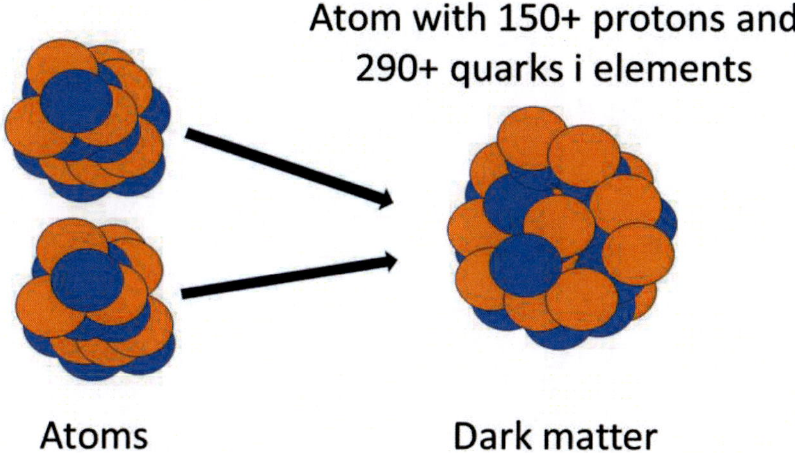

Figure 27. Two atoms fuse to form dark matter. (Credit: Bent Rolf Pettersen).

Chapter 27

Forces

Different energies possess distinct properties, which determine how they interact with other energies—whether similar or different.

A force arises from the interaction between energies.

Our physics is based on our universal energies and how they interact as individual energies and organized energy systems.

We now examine several of the primary material energies, their properties, and the forces they generate.

Interaction Matrix of Material Energies

	The strong nuclear force (YT energy)	Ilefos	Dielectricity	Uni energy (TE energy)
The strong nuclear force (YT energy)	Very strong attraction	Strong attraction	Moderate attraction	Strong attraction
Ilefos	Strong attraction	Strong attraction	Weak/moderate attraction	Strong repulsive reaction
Dielectricity	Moderate attraction	Weak/moderate attraction	Strong attraction	Strong repulsive reaction
Uni energy (TE energy)	Strong attraction	Strong repulsive reaction	Strong repulsive reaction	Extremely strong repulsive reaction

Our physical reality is largely shaped by these energetic interactions. Other energies may exist within organized energy systems, yet their influence is primarily internal and does not extend to interactions between separate systems. Dimensional energies are not included in the table, as they exhibit minimal interaction with the material energies listed.

The reactions create the forces we see in our daily life. They can explain atomic bonds between atoms, gravity, friction, energy transfer between atoms ("electricity"), and other behaviors in our universe.

Chapter 28

Stars

Stars are pure energy arranged an organized system. They consume energy in the form of matter, quarks, and other organized energy systems, as well as dark energy.

When matter and other organized energy systems are led into a star, the organized energy systems are dissolved and the energies are released. A star's primary energy source is the dissolution of atoms and other organized energy systems.

The intensity of a star depends on the availability and type of its energy supply. The types of energy and the amount determine which type of star we have.

If a star has access to dark matter, it becomes its most potent energy source. The second best is ordinary matter, and the third is other smaller quark groups. The amount of free energy units (dark energy) in the surroundings can also affect the star's properties. If the star is hit by pure energy beams from quasars or other sources, this will also provide a strong energy input.

- Dwarf stars have used up most of the fuel in their environment and have a low feed of matter/organized quark groups.
- Normal stars have a good feed of matter/organized quark groups from the surroundings.
- Massive stars, such as type O stars, have large amounts of matter to feed on, including dark matter.
- Quasars are sustained by vast amounts of energy originating from a black hole's Hawking radiation. A massive stream of energy and smaller quark groups are received directly from the black hole.

The core dynamic of a star involves the dissolution of matter and the reorganization organization of the released energies. Most of the energies are converted to dielectricity and ilefos. The star constantly releases large amounts of ilefos energy units, which form very powerful ilefos tracks (gravity tracks). These gravity tracks draw in a steady supply of matter from the surrounding environment. Ilefos attract dielectricity, and a large amount of dielectricity energy units are also carried along the ilefos tracks. Strong dielectricity energy is perceived as heat. The star also releases free dielectricity energy units that do not attach themselves to the ilefos tracks (gravity tracks). These can form small quark groups such as photons or resa. The star might also release other energies.

Stars consist of all the other energies stored in the quarks of organized energy systems, such as atoms, but most of the energy is converted to dielectricity, ilefos, and the YT energy (the strong nuclear force).

In the center of a star is a field of the strong nuclear force, YT energy. This energy attracts most other energies and helps the energies to remain in an organized structure.

The star also contains some dimensional energies that help organize the energies with the strong nuclear force (YT energy) and UR energy. The only form of matter-related energy the star does not retain is the repulsive energy (TE energy/uni energy). This energy is completely converted to dielectricity after being released from its quarks.

The Birth of Stars

When large quantities of matter are accumulate due to mutual gravitational attraction, a star can be formed. If the group of matter also includes dark matter, dark matter will be in the center of the group due to its strong gravity. This strong gravity will help draw other matter into the group.

The matter cluster can trigger a process of atomic or matte dissolution in two ways:

- Their inherent gravity (the attractive force as modeled by ilefos dynamics) can draw the atomic cores together to create a strong collision between the nuclei, creating a fission and an atom dissolution process that ignites the star.
- If a strong beam of energy hits a group of matter, this can also start a fission and atomic dissolution process.

When a star ignites, an atomic dissolution process begins. This process releases vast amounts of energy, coordinated by dimensional energies, YT energy (strong nuclear force), and UR energy. The strong nuclear force's attraction helps keep the energies together and organized in the star. As with atoms, a star can also transform energies. The freed energies that the star does not require are converted to the energies it does need. The star does not need repulsive force (TE energy) and certain other energies, and these are therefore transformed primarily into dielectricity, but also into ilefos and YT energy (the strong nuclear force).

The intensity of these processes depends on the star's energy input. If the energy input, for example, in the form of matter, is reduced, the processes in the star are also reduced. A strong white star type 0 might be reduced to a weaker type M star.

The Death of Stars

When the energy feed of a star is reduced, the processes and the star's energy are reduced—less energy results in a weaker nuclear force (YT energy) holding the star's energies together.

What happens next depends on the size of the star.

A massive star contains a lot of energy held in place by YT force (the strong nuclear force). It requires substantial YT energy to withhold all the other energies. As the energy processes in a star diminish, its YT energy is also reduced. If this falls below a threshold, the YT energies will not be able to hold back the star's energies, and it will release more energies, as its remaining energies will not be sufficient to maintain the star's internal processes. The internal system will collapse, and all energies held back by the YT energy are released in a supernova explosion. A supernova explosion releases a massive amount of energy in a short space of time. This energy might form quarks and resa, which through energy growth and fusion can also form matter.

The remnant core of the supernova may collapse into a small black hole, a neutron star, or remain as a residual star. The gases of the supernova explosion most often form a nebula, which in turn can grow into new stars.

A sudden energy increase to a white dwarf might also create a supernova explosion.

A medium or small star experiencing a significant decline in energy input will also have weaker internal energy processes. This results in a gradual reduction of the YT energy (strong nuclear force) holding back the energies in the star. The star's YT energy will be reduced, but it will not surpass the threshold at which all processes in the star stop. When the YT energy is slowly reduced, the star's density is reduced, and the star will grow in size. Its core will become colder, but the surface will remain warm. It will continue to grow due to the reduced attraction (YT energy/strong nuclear force) and become a red giant. Eventually, it will fade out and become a dwarf star.

A star might also gradually shrink in size and become a dwarf star.

Chapter 29

Black Holes

A black hole is defined as a region where gravity—modeled in this system as the attractive force of ilefos dynamics—is so strong that nothing can escape it. Neither light nor electromagnetic waves can escape from a black hole. The immense gravity of a massive black hole creates conditions conducive to galaxy formation. At the center of each galaxy, we find a massive black hole. The dynamics of galaxies—their gases and stars—depend on the gravitational pull of a central black hole.

Black holes can vary from a small stellar mass, formed when a massive star collapses, to supermassive black holes with millions of solar masses. At the center of our galaxy, the Milky Way, lies a black hole, Sagittarius A, which is a supermassive black hole of around 4.3 million solar masses in size.

The black hole of quasar J0529-4351 is estimated to be between 17 billion and 19 billion solar masses in size. This is the largest and fastest-growing black hole known to date. Black holes of this magnitude "eat" more than one sun a day, its quasar quasar is the brightest object currently known. It shines with a light that is 500 trillion times brighter than our sun. The light has travelled 12 billion years to reach us.

Black holes of various sizes can merge, forming even more massive black holes. When two galaxies collide, the attraction of the black holes at the center of the galaxies will also merge over time, forming a larger black hole.

Its immense gravitational force constantly draws matter and other organized energy systems into the black hole, including stars. Within it, all organized energy systems are dissolved, and the energy is released. Before matter crosses the event horizon, it orbits near the edge of the black hole. Here, we find a large amount of matter circling very rapidly and with very strong gravity. This intense friction generates heat and light in the surrounding accretion disk.

The currently closest identified black hole to Earth is Gaia BH1, which is around 1,560 light years away. Its size is around 9.62 M⊙ (9.62 solar masses).

The largest black hole found in our galaxy, if we disregard Sagittarius A at the center of our galaxy, is located in the constellation Aquila. This black hole is 2,000 light years away and spans around 33 M⊙ (solar masses). (Gaia

Collaboration: P. Panuzzo, T. Mazeh, ... "Discovery of a dormant 33 solar-mass black hole in pre-release Gaia astrometry," Astronomy & Astrophysics journal, 2024).

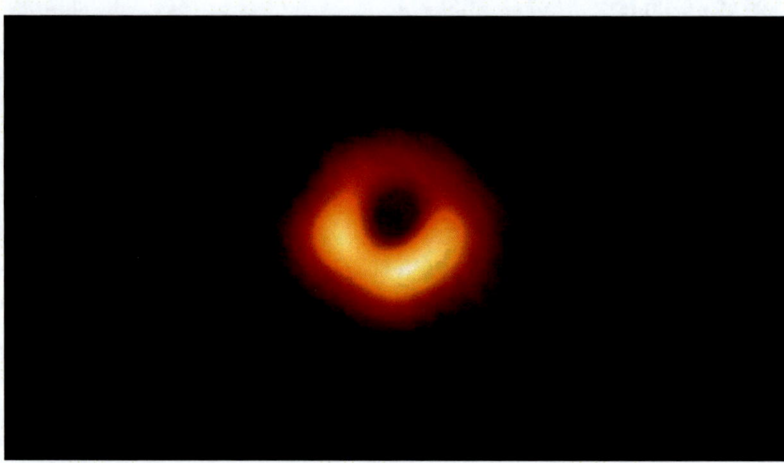

Figure 28. The first image of a black hole in the center of the M87 galaxy. Credit Event Horizon Telescope Collaboration.

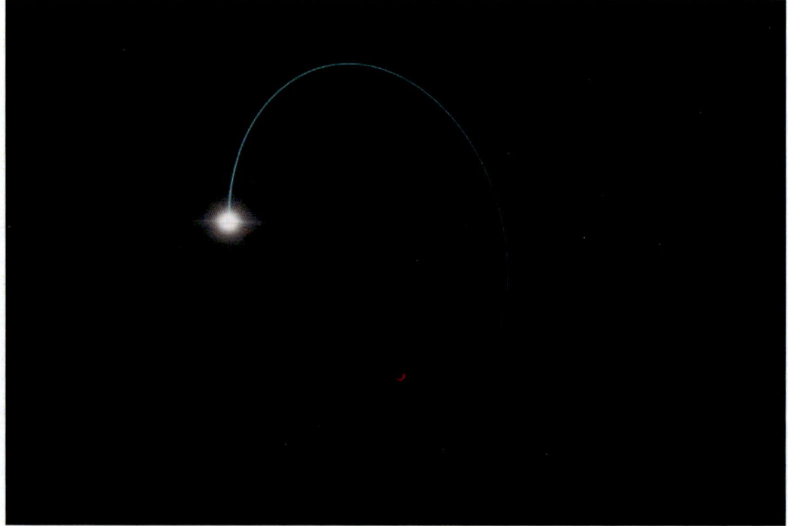

Figure 29. A star in orbit around a 33 solar mass black hole in the constellation Aquila, 2,000 light years away, which in turn orbits around Sagittarius A, the black hole at the center of our galaxy, in one spiral arm of the Milky Way. (Artist impression. Credit: ESO/L. Calçada).

Chapter 30

The Dynamics of Black Holes

What are the physics of a black hole?

A massive black hole is fed by matter and other organized energy systems, including pure energy systems like stars, and dark energy.

Smaller black holes primarily receive their energy from matter.

The material feeding a black hole first reaches a zone around the black hole with very strong gravity (the attractive force as modeled by ilefos dynamics), where the matter circles rapidly before entering the hole. This friction generates heat and light in the accretion disk surrounding the black hole.

After this, the material enters the domain of the black hole. Here, the very strong gravity makes the cores of the organized energy systems collide in strong collisions that destroy their atomic elements and quarks. The energy plasma of stars is already organized in streams of pure energy, which become part of the black hole's administration system.

Within a black hole, all organized energy systems are transformed into raw energy. It is therefore made up of the same energies we find in organized energy systems like matter and stars, but they are organized differently.

In stars, the strong nuclear force can organize the material energies. Assuming that a black hole consists of a concentration of all universal energies, it would require a universal energy capable of interacting with all energy types, not only material energies. Dimensional energies and energies unaffected by the strong nuclear force (YT energy) also need to be addressed. As such, we must assume that a black hole requires a special energy that can also attract these energies for the black hole to organize them. I call this energy UR energy.

A black hole has two primary energies that help it organize its energies:

- YT energy (the strong nuclear force)
- UR energy

YT energy is the same energy that allows matter and other organized energy systems, like stars, to organize their energies. This energy exerts attraction on all material energy forms. It is the main energy required to organize ilefos energy (the attractive force, gravity) and dielectricity. Strong YT energy withholds most of these energies in the black hole system. The energy also organizes other material energies with an attraction towards YT energy.

UR energy exerts attraction across all energy domains, including dimensional energies. This energy organizes YT energy (the strong nuclear force), and the dimensional energies that carry and organize information. UR energy may be considered the first foundational energy of the universe, and it can interact with all other energies. Dimensional energies, together with UR energy, create black holes and the universe.

The black hole requires a constant energy feed, mainly from organized energy systems. To secure a steady flow of matter, the black hole must send out a stream of Ilefos energy units. These units self-attract, forming massive ilefos tracks (gravity tracks), which attract all organized energy systems such as matter and stars. The attraction causes them to accumulate around the black hole, often in spiral arms. These gravity tracks are extremely strong and attract far beyond the immediate area of the black hole. The Ilefos tracks may therefore link to gravity tracks from other galaxies and with external gravity fields.

According to the Ilefos model, approximately 98% of a black hole's incoming energy is re-emitted from the black hole through gravity tracks (Ilefos tracks) and Hawking radiation. With this dynamic, a black hole constantly grows with two percent of its energy input.

The Ilefos energy units of Ilefos tracks (gravity tracks) have an attractive force on each other. Therefore, an Ilefos track attracts other Ilefos tracks, forming a stronger gravity concentration around the black hole's spin area. We also see this effect in stars where the Ilefos tracks' mutual attraction forms a stronger gravity in the direction of spin, where we find the planets in orbit. Around the black hole, the mutual attraction of the ilefos tracks intensifies gravitational fields around the black hole's spin axis. This stronger attraction creates the spiral arms of a spiral galaxy.

As with matter and stars, a black hole also appears to be an organized energy system, but the internal energy concentration seems extremely strong. This concentration creates an intense attraction that does not allow dielectricity and other energy to escape the black hole. Heat and photons require dielectricity. Because the black hole's gravity retains dielectric energy, the black hole does not emit heat (dielectricity tracks) or photons. The exception is directly upwards and downwards from the direction of spin. Here, the gravity is weaker, allowing some dielectricity and other energy units to escape. These energy units may coalesce into photons, minor quark aggregates, or light gases. This is Hawking radiation. These quark groups also attract some gases from the accretion disk around the black hole.

Chapter 31

How a Black Hole Grows

Black holes grow through two main mechanisms: energy growth and fusion:

- Energy growth
- Fusion

Energy Growth

A black hole constantly feeds on matter and energies pulled in from its surroundings due to its strong ilefos tracks (gravity tracks). Organized energy systems are dissolved, releasing their component energies. These energies are reorganized within the black hole by its internal administration system.

A black hole releases a large quantity of ilefos energy units, which form strong ilefos tracks when emitted from the black hole. Larger black holes also release dielectricity and other energies through Hawking radiation from the "top and bottom" of the black hole. These energy units might form quarks and simple, organized energy systems. Hawking radiation also carries some material from the accretion disk. A black hole emits energies by gravitation and Hawking radiation, but not all energy input is emitted this way. Around two percent of the energy input is kept in the black holes to preserve their structure. This allows the black hole to grow by energy. It requires energy to manage internal organization and perform its functions, including those involving dimensional functions. Therefore, it is essential for the black hole to retain some of the received energies.

If black holes encounter areas with strong dark energy, such as dark energy fields, the black hole absorbs the increased dark energy and can result in additional energy growth.

Energy Growth by Fusion

The immense gravity tracks may attract those of another black hole. This can draw two black holes toward each other and cause them to spin around each other until gravitational forces cause them to merge into a black hole fusion. When two black holes merge, the operating system of the largest black hole takes over the administration of the energies. The smallest operating system dissolves, and all energies from this black hole lack organization, for a short time, resulting in an extremely dense concentration of unbound energies. The majority of these energies are absorbed by the largest black hole.

The smallest black hole's operating system ceases to exist, and all organization of energies ceases. When the strong nuclear force (YT energy) and UR energy retaining the energies in a system stop their functions, there is nothing to hold the energies back. Therefore, all energies are emitted in an immense energy release. Most of these energies are absorbed by the largest black hole, but around 40 percent of the energies released are not. This includes strong emissions of ilefos energy units—perceived as gravitational waves, which were detected on Earth by LIGO gravitational wave detectors in 2015.

When two black holes merge by fusion, the new black hole has all the energy of the largest black hole and around 60 percent of the energy of the smallest black hole that was dissolved. Thus, black holes also grow through fusion.

If two galaxies collide, the black holes at the center of each galaxy will, over time, merge by fusion. The released energy of the dissolved black hole results in increased energy, which forms matter and can create new stars.

Chapter 32

Our Visible Universe

What we observe through telescopes constitutes our visible universe — composed of energies and energy systems formed by our Big Bang. These are energies and energy systems formed by our Big Bang.

Beyond our visible universe lies space governed by the same physical laws. This broader domain is referred to as the grand universe. The grand universe contains vast amounts of energy, primarily in the form of free energy units —dark energy. These energies predate our Big Bang.

The grand universe is enormous, many times larger than our visible universe. Because few organized energy systems exist there, this domain remains mostly calm, with little interference affecting the free energy units (dark energy). As a result, similar free energy units —being mutually attractive—may form large, dense energy fields of free energy units.

Ilefos energy units exhibit strong mutual attraction. This leads to the formation of dense regions of ilefos energy units and gravity fields. Large gravitational fields may thus emerge in calm zones beyond the expanding domain of our Big Bang.

In addition, other energies —such as dielectricity, YT energy (the strong nuclear force), and dimensional energies —may generate distinct field structures in calm areas of the grand universe.

Chapter 33

The Accelerating Expansion of the Visible Universe

Our visible Big Bang universe expands with an accelerating speed higher than the speed of light.

In general relativity (Albert Einstein, “Die grundlage der allgemeinen relativitatstheorie,” 1916), this is explained as an expansion of the fabric of time and space. We do not move; it is the fabric of space that expands. This seeks to explain how the universe expands with an accelerating speed higher than the speed of light.

However, this theory has weaknesses. The nature of the space-time fabric remains unexplained. The theory is based on the expansion of a fabric nobody can explain. And if the fabric of space expands, all space should expand. Yet we see clusters of galaxies and matter in which space does not expand. Galaxies and areas of matter seem to be mosaics of static space within an an expanding universe. This might indicate that general relativity is not a correct description of the expanding universe.

The Inflation Theory proposes that the universe began as a sudden expansion after the Big Bang. Cosmic expansion then decelerated to much slower speeds. Around 9.8 billion years after the Big Bang, around four billion years ago, the universe gradually expanded more quickly. The universe has continued to expand at an accelerating speed and now expands at a speed higher than the speed of light.

Scientists have explained this by the existence of dark energy, which also appears in the cosmological constant. The Lambda-CDM model, Lambda cold dark matter, is a mathematical model that explains the cosmological constant with dark energy. This is considered the standard model of Big Bang cosmology. Dark energy is an unknown form of energy that affects the dynamics of the universe and drives the accelerating expansion.

In this chapter, I will explain the accelerating expansion of the universe in accordance with the Ilefos model. The Ilefos model explains the basic material energies and how they interact with each other. In this model, dark energy consists free energy units of material energy not bound in organized systems like matter, stars, and black holes.

The speed of expansion depends on the measurement point. If we measure the speed of our current expansion from the ground zero of the Big Bang, the expansion is now estimated to be 3.5 c (the speed of light). The relative recession speed compared to a location on the opposite side of the visible universe, which moves in the opposite direction, is 7c.

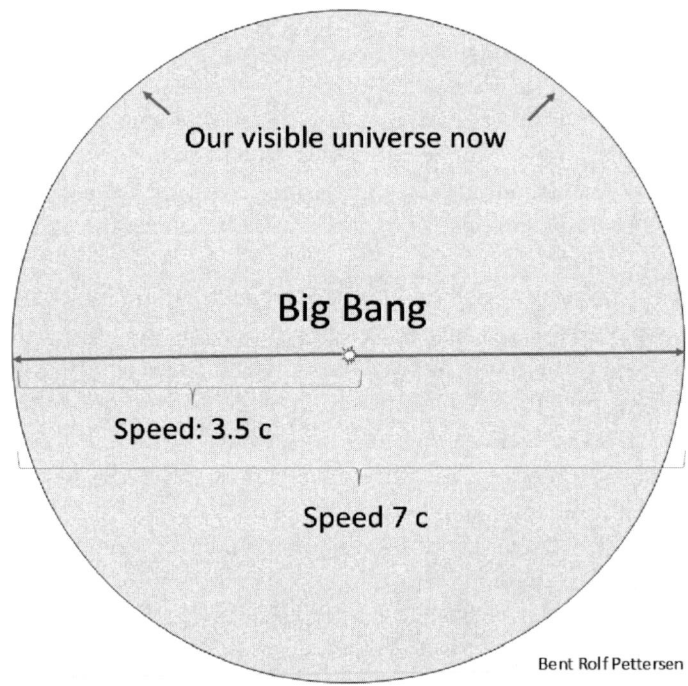

Figure 30. Speed of expansion of the visible universe according to the Ilefos model. From the Big Bang center, the expansion is 3.5c. If we measure the speed from a point on the other side of our visible universe, we move away from this point at a speed of 7c. (Credit: Bent Rolf Pettersen).

The Expansion Speed of the Universe

The Hubble–Lemaître law (Hubble's law) states that galaxies move away from Earth at speeds proportional to their distance. (Edwin Hubble, "A Relation Between Distance and Radial Velocity among Extra-Galactic Nebulae," Proceedings of the National Academy of Sciences, 1929). The farther away they are, the faster they move away from Earth. The velocity of galaxies is determined by the shift of light they emit towards the red end of the visible spectrum (redshift).

The Hubble constant tells us how fast the universe is expanding and indicates that the expansion rate should be around 68 km/s/Mpc. A megaparsec (Mpc) is one million parsecs or 3,260,000 light years. The European Space Agency's Planck satellite gave us a Hubble constant of 67 km/s/Mpc based on time fluctuations of the cosmic microwave background radiation.

Cepheid variable methods use pulsating, aging stars to measure distances. According to Cepheid variables, the universe's expansion rate is 74 km/s/Mpc, a much higher value than Planck's measurements.

In 2023, more accurate data were collected from the James Webb Space Telescope. Astronomers found the new observations to confirm the Cepheid variables of the Hubble telescope. The EPFL study of 2023 (Mauricion Cruz Reyes, Richard I. Anderson, "A 0.9% calibration of the Galactic Cepheid luminosity scale based on Gaia DR3 data of open clusters and Cepheids," Astronomy & Astrophysics, 2023) seems to provide a precise expansion rate of 73 km/s/Mpc.

According to Professor Richard I. Anderson of EPFL's Institute of Physics, "The discrepancy has many other implications. It calls into question the very fundamentals, like the exact nature of dark energy, the time-space continuum, and gravity. It means we have to rethink the basic concepts that form the foundation of our overall understanding of physics."

Based on this study, the universe's expansion rate is estimated at 73 km/s/Mpc.

The Hubble–Lemaître law (Hubble's law) is based on Albert Einstein's general relativity, which states that the fabric of space and time expands. As mentioned earlier, however, this theory has several weaknesses, indicating that it could be wrong. If we instead interpret the expansion as motion through a pre-existing space, then the Big Bang universe would effectively recede at superluminal speed relative to distant points. To explain this, there must be a force to push or pull the universe. Our current explanation is that something pushes the universe, but in order for something to accelerate, the acting force must increase. Scientists have struggled to understand how a pushing force might constantly increase.

If we have a force pulling the universe, this force will increase as the distance between the source of this pull and the universe decreases. This could result in an accelerating and expanding speed. In the next chapter, we will take a closer look at dark energy fields outside our visible universe and how they might explain the expanding universe.

Dark Energy Fields – Gravity Fields

Outside our visible universe, in the grand universe, we can assume that there is as much energy as in the visible Big Bang universe. The energies outside our visible universe consist mostly of free energy units, not concentrated in organized energy systems like matter and galaxies. The grand universe, therefore, consists primarily of dark energy.

A large portion of this energy is the same energy we find in matter and has the same properties as matter.

Ilefos energy units have a strong attraction towards other ilefos energy units. In calm areas of the universe not affected by gravity, tracks from organized energy systems, this attraction makes the ilefos energy units form dense areas of free ilefos energy. This is gravity fields (ilefos fields). Ilefos energy units might form in extremely large gravity fields that surround the visible universe.

These gravity fields have immense gravitational force. When the expansion of the visible universe started to increase in speed around 9.8 billion years after the Big Bang, this was caused, according to the ilefos model, by the attraction of these gravity fields. This attraction became stronger than the mutual attraction of matter in the visible universe (Big Bang universe), which until that point had started to slow down the expansion.

In order for something to increase in speed, the acting attractive force must increase in magnitude. With a magnet, the attractive force increases when we decrease the distance between the magnet and the magnetic material. In the same way, the visible universe experiences a stronger attractive force when the distance between the materials in the visible universe and the energy fields decreases. The attraction of gravity fields pulls the expanding universe with an increasingly stronger attractive force, causing the universe to expand at an accelerating speed.

As previously discussed, the attractive force mediated by ilefos energy propagates far faster than the speed of light. Therefore, the attraction makes our visible Big Bang universe expand at a speed higher than the speed of light.

Chapter 34

The Universe

Our universe is defined as a domain filled with our universal energies and governed by our specific physical laws. These energies may exist as free energy units (dark energy) or as organized energy systems such as quark groups, matter, stars, and black holes.

Other external universes may be governed by different energy forms and alternative physical laws.

Our universe, the grand universe, is an area with our energies and ruled by our physics.

Our universe can be conceptually divided into two parts: the visible universe and the grand universe.

- The visible universe (materials from our Big Bang)
- The grand universe (the universe outside the visible universe)

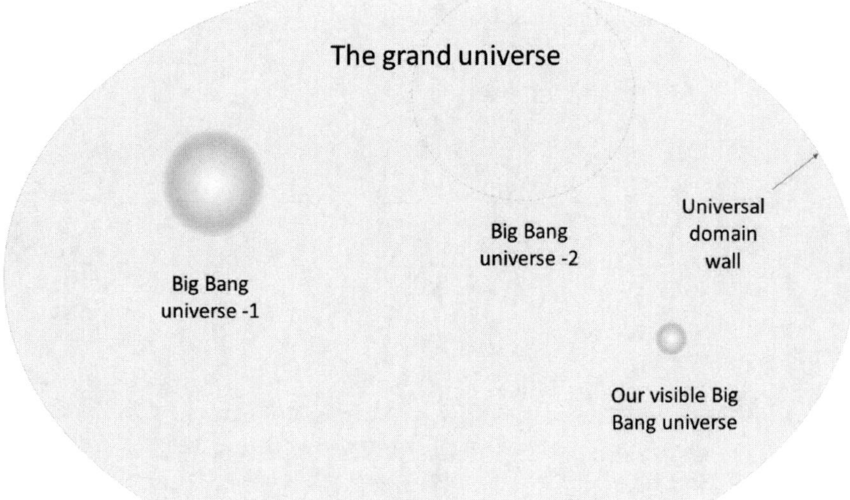

Figure 31. The grand universe. (The universal domain wall will be explained in a separate chapter.) (Credit Bent Rolf Pettersen).

The visible universe refers to the region observable through telescopes and other instruments. This universe is where we find a large proportion of organized energy systems.

The grand universe is dominated by free energy units —referred to as dark energy. It is enormous, much larger than the visible universe. We can also find organized energy systems in the grand universe, but it is dominated by free energy units. Organized energy systems in the grand universe are leftovers from old Big Bangs in other areas of the grand universe.

Chapter 35

Energy Distribution in the Universe

The universe consists of energy fields—either concentrated in organized systems or existing as free energy units.

Energy units within these fields are attracted to one another based on their inherent properties.

The exception is TE energy (uni energy, the repulsive force), which does not form separate TE energy fields. TE energy repels all other energies except for YT energy (the strong nuclear force) and UR energy. Free TE energy units may bind to YT energy fields as auxiliary components.

All other energies have an attraction towards equal energy units and form separate energy fields in calm environments, without disturbance from matter and organized energy systems with ilefos tracks (gravity). The areas between these energy fields are dominated by dark energy, formed of free energy units from different energies not concentrated in separate energy fields.

Regions with high matter concentration tend to exhibit a predominance of dark energy.

Matter absorbs significant energy when interacting with energy fields in the grand universe. This creates dark matter, which grows to maximum strength and size and can be absorbed by stars and black holes. This can lead to explosive growth in the size of black holes. These black holes can fuse due to the immense gravity to new, larger black holes.

A black hole can then rapidly reach its saturation point. At this point, the black hole can no longer maintain structural coherence due to an energy overload. The organized energy system of the black hole will then dissolve, and all energies will be released in a big bang. An extreme release of all energy types follows, manifesting as dark energy. The concentrated energy after a big bang creates a substantial amount of matter and dark matter, which in turn creates stars and black holes. Energy units are recycled and can be converted into new energies in organized energy systems. This is the dynamics of the universe in the Ilefos model.

Our visible universe has a high proportion of organized energy systems, such as quark groups, matter, dark matter, stars, and black holes.

The grand universe beyond our visible domain is composed almost entirely of dark energy (free energy units), which might form fields of specific free energy units (energy fields). The grand universe also contains some organized energy systems, which are leftovers from earlier big bangs that a universal domain wall has not yet dissolved. (Universal domain wall will be explained in a separate chapter).

Energy in the universe exists either in organized systems or as free energy units (dark energy), which might form energy fields (dense fields of energy units), which may aggregate into energy fields.

Chapter 36

Energy Balance in the Universe

The universe —composed of the visible and grand universe —was formed with a fundamental level of universal energy. The properties of these energies, and their interactions, define our physics.

Over time, these energies interacted and formed organized energy systems based on universal laws. The "Atomic DNA," the blueprint of organized quark groups, gave us matter and dark matter, which in turn became stars and black holes, and finally contributed to a Big Bang. The blueprint of organized energy systems must rest upon a universal dimensional energy that can interact with all universal energies. This dimensional energy transmits information and instructions to all universal energies and can be regarded a universal administrator energy.

The energies of our universe are contained by universal domain walls. These boundaries retain our universal energies and shield the universe from external energies with incompatible properties that might disturb our universal system.

The universe is in energy balance. The universal domain walls retrain the universal energies. Only select multiversal dimensional energies are permitted to exit and enter. These represent energies shared by all universes, regardless of their unique energy types or governing rules.

Chapter 37

The Universal Domain Wall

Free energy units and organized energy systems travel through the universe. The universal domain wall marks the boundary of our universe. Our universe is characterized by our energies and their respective properties. These properties define the physics of our universe. The universal domain wall is the border for the area of our energies and physics. Outside the universal domain wall, we find other universes with different energies and physics.

To retain energies within our universe, free energy units that reach the universal domain wall are reflected back into our grand universe. Organized energy systems reaching the universal domain wall are dissolved, the energies are freed, and their energies are returned to the grand universe as free energy units (dark energy).

The universal domain wall dissolves organized energy systems, and the grand universe therefore has more free energy units (dark energy) than in our visible universe. The domain wall prevents all universal energies from escaping our universe, except for specific multiverse-dimensional energies shared among all universes. These enable inter-universal information exchange.

The universal domain wall is essential to our universe to protect our physics and universal dynamics. It is a membrane that only allows multiverse-dimensional energies to pass. All other energies, whether free energy units or organized energy systems, are sent back into our universe as free energy units.

The universal domain wall has two main purposes:

- It retains the energies in the universe
- It protects the universe from external energies not part of our universal system. External energies might dynamically react to our universal energies and destroy our universal system.

In the grand universe, the mutual attraction among similar free energy units gives rise to expansive energy fields, dense with specific energy units, such as ilefos energy fields (gravity fields) that are dense with ilefos energy units, or dielectricity energy fields with dielectricity energy units. All universal energies, except TE energy (the repulsive force), create energy fields in calm areas of the grand universe.

The universal domain wall establishes stability and energy balance across the universe, which enables energy fields and organized energy systems to be formed.

Chapter 38

UR Energy

The idea of Grand Unification has been around for around 50 years. Scientists believe that all three forces originate from the same "energy" at the beginning of the universe. This means electromagnetism (including the electroweak force), the strong nuclear force, and gravitation originate from a unified force, or a unified state, which existed early in the history of the universe. This is the Grand Unified Theory, proposed by Howard Georgi and Sheldon Glashow in 1974 (Georgi, Howard; Glashow, S. L. "Unity of All Elementary-Particle Forces," Physical Review Letters, vol. 32, Issue 8, pp. 438-441, February 1974). At sufficiently high energy levels, the fundamental forces unify into a single force.

In order to manage all energies in the ilefos model, we must introduce an energy that has an attractive force towards all energies, both material and dimensional energies. This energy can thus organize all other energies forms in the universe. I call this energy UR energy, the energy of the unified force. This energy must then also appear in lower energy environments. Since all energies originate from it, UR energy can attract all other energies.

Assuming this is the case, it must also be able to organize all energies. It can be converted into, and can itself convert, all other energies in the universe, and it can convert all other energies. The neutral quark in atoms has some UR energy, which enables it to convert energies during Atomic Phase Displacement. This is observed in iron, which can convert dielectricity to ilefos (magnetism) in an electromagnet. The neutral quark with UR energy in copper can convert ilefos (magnetism) to dielectricity, as seen in a generator. An atom can convert most forms of energy via the UR energy in its neutral quark.

We find concentrated UR energy in the grand universe's dark matter, stars, black holes, and special energy fields.

In larger organized energy systems, on a greater scale than atoms, UR energy helps to organize the energies. It is much stronger than the YT energy (the strong nuclear force) and can therefore manage extensive and strong energy systems, such as those we see in dark matter and black holes.

UR energy can organize, concentrate, and convert all other universal energies. It must therefore be the first and most substantial energy; conversions of UR energy created all other energies in the universe.

UR energy governs the universe in conjunction with ER energy—a dimensional energy of information and communication. ER energy is a dimensional energy of information and communication. UR and ER energies may be considered the primordial energies—and organizing forces—of our grand universe.

Chapter 39

Critical Mass of a Black Hole

A black hole dissolves organized energy systems, releasing their constituent energies. The energies released, which are different types of energies, are then controlled and organized by the black hole. Most of these are material energies, attracted to YT energy (the strong nuclear force). The operating system of a black hole must be able to administer YT energy (the strong nuclear force) in order to organize the energies into a system.

The black hole also has to manage dimensional energies and other energies not attracted by YT energy. UR energy is a universal energy that can organize all energies. UR energy can attract all energies and can be converted to all other energies. This is the main universal energy, together with some dimensional energies.

Inside a black hole we find a very dense plasma of different energies. Different energy types form distinct energy layers or fields. YT energy (the strong nuclear force) creates a layer of attractive energy that retains most of the material energies.

As with atoms, black holes also seem to have an advanced operating system that administers all tasks and energies. A large percentage of the received energies are converted into ilefos energy (the attractive force), which in is then emitted at high speed from the black hole. The ilefos energy units have an attractive force toward each other and form strong bonds of ilefos energy units and gravity tracks when emitted from the black hole.

These very strong gravity tracks span long distances, far beyond the galaxy, and can also attract external organized energy systems at great distances.

The gravity tracks of the black hole draw in matter and other energies. The tracks constantly feed the black hole with energy, and it must constantly emit large amounts of ilefos (gravity tracks) to operate. Approximately 98% of the energy received by a black hole is assumed to be re-emitted as ilefos (gravity) tracks to ensure a constant steady influx of energies.

Some energies are also released in areas of lower gravity, such as Hawking radiation, but this is only a small amount of the energy input. Around

1.5 to 2 percent of the received energies are retained in the black hole contributing to its internal organization and mass increase.

The black hole then grows constantly if it is fed. The layer of YT energy (strong nuclear force) and UR energy retain and control all of the energies of the black hole. A black hole can grow to an enormous size if it is fed with a lot of energy over a long time. Also, the fusion of black holes results in larger black holes.

A black hole increases in size at a somewhat steady rate. When black holes from an expanding Big Bang universe approach an ilefos energy field (gravity field) of the grand universe, the attraction between the gravity field and the expanding universe causes the speed of the expanding universe to increase. Finally, the expanding universe will come into contact with the energy fields of the grand universe. All organized energy systems will receive substantial energy and exponentially grow in size/energy. Matter will become dark matter and grow to its maximum size. The black holes will then also exponentially grow in size.

Many black holes will fuse due to their increased attraction, and the black holes can become enormously large.

With increased energy, the black hole's rotation speed will also increase. An enormous mass creates an enormous rotation speed. A black hole approaching Big Bang scale will eventually rotate near the speed of light. When this happens, the centrifugal force will change the shape of the black hole to an oval. The stress of the boundary further increasing stress on the event horizon. The UR and YT energy will eventually be unable to retain the ilefos and other energies.

The YT energy and UR energy, which hold back and control the energies of the black hole, will at that point be unable to handle the explosive increase of energy and the increased centrifugal force. The black hole then reaches its energy handling threshold and will lose control over its energies. This extreme black hole now contains a Big Bang-scale energy system, which it cannot control. All of these energies are then released as free energy units in an explosion of energy. This is a big bang.

The big bang-sized black hole has an oval shape just before it explodes. With the immense centrifugal force, the shape will also influence the explosion and the release of energies.

Gravitational strength is reduced along the spin axis—upward and downward—making it the initial axis of release. The energy release will start here, followed by an immense release of energy in all directions, but predominantly outward along the equator, where centrifugal force and mass concentration are highest. The nature of this black hole explosion also forms the subsequent Big Bang universe, like our visible expanding universe.

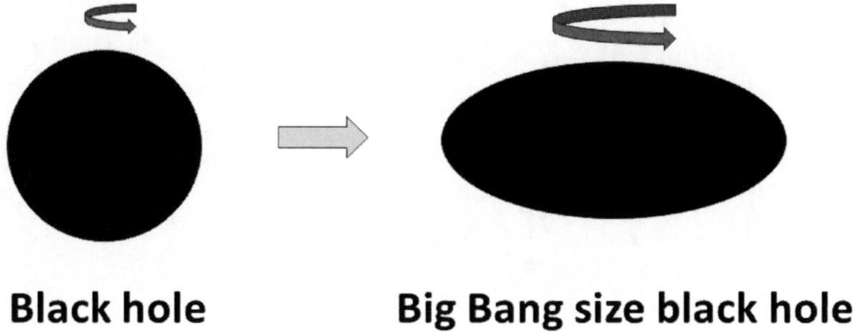

Figure 32. With increased mass and spin, a black hole close to critical mass takes on an oval form. (Credit: Bent Rolf Pettersen).

This event can occur in different places in the grand universe, but the time interval between occurrences of a Big Bang is extremely long.

Chapter 40

A Big Bang

As a black hole approaches Big Bang-scale mass, its internal dynamics begin to shift. The immense energies within the black holes are structured in layered energy fields. The UR energy organizes and manages most of the energies. UR energy, being attracted to all other energies, can thus calibrate and manage the various energy fields in the black hole.

The increasing mass causes the black hole to spin faster and faster, to a speed at which the centrifugal force presses the energy fields outward with immense force. The centrifugal force then starts to affect the UR energy, which also moves outwards.

This distortion alters the black hole's shape into an oval. The rotation speed continues to increase, and the oval shape makes the centrifugal force at the outer edge of the black holes even greater. The UR energy has to manage its energy and all the other energies in what is now an extreme gravitational force.

When UR energy can no longer contain the outer energy layers of the spinning black hole, the black hole has reached its threshold. Unable to retain control, the organizing UR energy releases the accumulated energies in a vast explosion. This release constitutes a Big Bang.

The energy release during a Big Bang is highly anisotropic. The oval shape of the spinning black hole releases most of the energy directly outwards in the direction of rotation. Gravitational pull is weakest along the spin axis—directly above and below, directly upwards and downwards. This is where some energies are released in Hawking radiation. When the black hole dissolves, we will also have a strong release of UR energy and some upward and downward release of other energies, yet the majority of the energies are released outward along the equatorial spin plane, where the gravitational forces are strongest.

A black hole with a mass approaching big bang magnitude that reaches its threshold and releases its energies is a big bang.

Chapter 41

The Shape of the Universe

If most of our visible universe consists of remnants from a big bang caused by an exploding black hole, the shape of our visible universe must be shaped by the nature of this explosion.

According to the theory of general relativity (Albert Einstein), the universe could take on three forms: flat, closed like a sphere, or open like a saddle.

The cosmic microwave background (CMB) is microwave radiation that fills all space in the observable universe. This radiation was discovered by American radio astronomers in 1965 (Arno Penzias/Robert Wilson, "Measurement of Excess Antenna Temperature at 4080 Ms/s," The Astrophysical Journal, 142, 1965). CMB showed us the faint background glow, which is not associated with stars, galaxies, or matter.

Several satellite and balloon experiments seemed to confirm the CMB, but with low resolution. The Planck space telescope, launched by ESA (European Space Agency) in 2009, gave us a high-sensitivity map of the CMB radiation, which improved the earlier observations made by telescopes and the WMAP (Wilkinson Microwave Anisotropy Probe/WMAP, NASA).

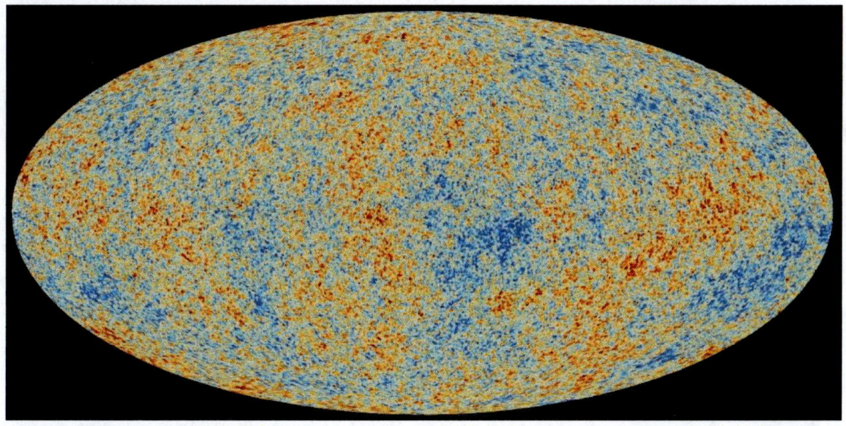

Figure 33. Cosmic microwave background, CMB, from the ESA's Planck mission. (Credit: ESA).

Astronomers have now measured the temperature fluctuations of the CMB very accurately. The results show the universe's density with a high degree of accuracy. Most of these results show that the universe expands in every direction without positive or negative curvature. "The universe is flat." This is the major component of the standard cosmological model (The Lambda-CDM model). The Lambda Cold Dark Matter (ΛCDM) model is the prevailing mathematical framework of the Big Bang theory, which assumes that general relativity is correct. The model assumes that the universe's shape is flat (zero curvature).

However, some studies show that if we take into account how much the gravity of matter distorts the CMB, is it more consistent with a closed universe, or a sphere (Eleonora Di Valentino; Alessandro Melchiorri; Joseph Silk (November 4, 2019). "Planck evidence for a closed Universe and a possible crisis for cosmology." Nature Astronomy. 4 (2): 196–203).

The CMB gives us a mixed description of the shape of the universe, depending on which theories and studies we follow.

To explain the shape of the universe, we must take a look at the dynamics of the Big Bang. In the Ilefos model, the Big Bang is an exploding, spinning, massive black hole that has reached the threshold for how much energy it can retain.

How will an exploding black hole behave? We know by the shape of galaxies and images of black holes (Event Horizon Telescope Collaboration, 2019) that the spinning black hole in the center of galaxies causes stronger gravity at an angle of 90 degrees to the spin axis. This is why spiral axis galaxies concentrate the attraction of matter and stars in one plane. The dynamics of a spinning object generate a centrifugal force that pushes mass outward along the plane of rotation. In a black hole, we must assume that the energy plasmas —concentrated layers of pure energy—are also denser along this plane. The gravitation is weakest in the axis, "directly upwards and downwards," where some energy that is not ilefos (gravity) escapes as Hawking radiation.

When a Big Bang-size black hole explodes, it releases some matter where the gravity is weakest, along the axis of the spinning black hole. This gives us a small peak of energy straight upwards and downwards. The bulk of this energy will is released where the energy plasma is most concentrated—directly outward along the equatorial plane (90 degrees from the spin axis). The exploding black hole will release its energies in all directions, but more in the radiant direction straight outwards.

Prior to explosion, the spin velocity likely approaches the speed of light, thus giving the black hole a slightly oval shape just before it explodes.

This explanation would imply that the visible universe has a unique shape: an elongated spheroid with two polar protrusions.

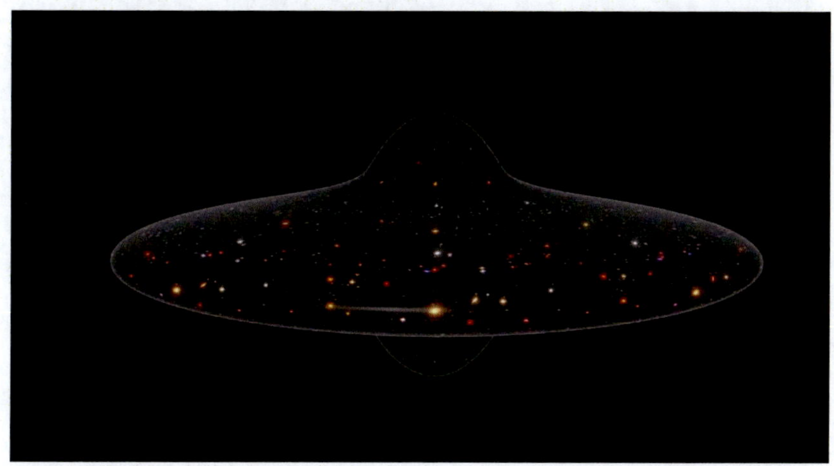

Figure 34. Shape of the expanding universe in the Ilefos model. (The visible universe contains much more mass than the illustration suggests.) (Credit Ola Tandstad).

A vast amount of matter is created in the first seconds after a big bang. This matter moves outwards from a big bang in a dense field, which may resemble a toroidal (donut-like) structure, resulting from the intense pre-Big Bang rotation, most energies that are affected by gravity experience a centrifugal force. When the black hole can no longer hold back the energies, most of these energies are released directly outwards in the direction of centrifugal force.

When released, energies might be organized, and systems like matter are formed. Most of the energies continue to travel outwards in a donut-shaped concentration of organized energies. At the same time, "lighter" energies — such as dimensional energies—escape along the spin axis, where centrifugal force is minimal. "Lighter" energies, such as dimensional energies, are less affected by gravity, but some normal material energies are also affected. Therefore, the visible material universe takes on the following shape:

Figure 35. Expanding universe in a donut shape with peaks. (Credit: Ola Tandstad).

It is highly probable that the expanding universe after a big bang takes on a shape resembling one of the two illustrations in this chapter.

These shapes are based on the dynamics of an exploding black hole. This is an alternative conceptual theoretical model that must be proven in the form of a mathematical model.

Chapter 42

The Nature of a Universe

Our universe is governed by a distinct set of energies. This energy composition and the nature of some of these energies are unique to our universe. The properties of these energies —and how they interact —define our universal physics.

Dimensional energies carry information and an intelligence that can set rules for organizing material and dimensional energies. The "blueprint" for energy behavior and organization in energy systems is transported by dimensional energies to all parts of the universe. As such, all organized energy systems have a point of contact with a dimensional energy.

In the Ilefos model, all material energy systems are managed by an atomic quark. This quark contains a dimensional energy that interfaces with the universal dimensional energy. The atomic quark is then given the behavior instructions for performing atomic actions and forms a material energy DNA guided by the universal dimensional energy. This allows energies to be organized in different quarks, creating organized energy systems such as resa/particles, matter, and dark matter.

Energies can also be organized in pure energy systems with fields of energies and pure energy plasma. When a star or black hole dissolves material energy systems like matter and dark matter, these energies are released and restructured into pure energy fields. We must also assume that we have other organized energy systems in our grand universe outside the visible universe.

Our grand universe must be separated from other universes by a universal domain wall. The universal domain wall dissolves all organized energy systems and sends the energies back to our grand universe. Free energy units (dark energy) are likewise reflected back into our grand universe. In this way, our grand universe maintains an energy balance.

Matter and large-scale phenomena emit attractive energy units, forming gravity tracks that weaken with distance from their source and will, in time, dissolve to form pure ilefos (attractive) energy units. These free energy units are part of dark energy, which we find throughout our universe. To be in energy balance, atoms must receive a constant flow of energy to replace the energy they release to form gravity tracks and other energy releases. Atoms receive a continuous inflow of dark energy to their nucleus via harvesting tracks. In this way, free energy units (dark energy) are recycled through atoms.

Energy in our universe is unevenly distributed. They may be concentrated in energy fields and organized energy systems. Organized energy systems, such as black holes, may grow to enormous sizes, at this point, they cannot withhold their energies and explode in big bangs. Big bangs distribute energy and give rise to new organized systems.

The properties of our energies decide how an energy unit reacts to similar or different energy units. The reaction between similar or different energies creates our physics.

Our energies and their properties create our universal physics. Other universes have different energies and rules of physics.

Chapter 43

The Beginning of the Universe

The Grand Unified Theory posits that all energies originate from a single unified force. (Georgi, Howard; Glashow, S. L. "Unity of All Elementary-Particle Forces," Physical Review Letters, vol. 32, Issue 8, pp. 438-441, February 1974).

In the Ilefos model, this unified energy is referred to as UR energy.

At the origin point of the universe, only UR energy and a dimensional energy were present.

I refer to this dimensional energy as ER energy, which is responsible for communication and organization. This energy can organize UR energy and allows UR energy to convert to other energies. ER energy encodes the properties of all energies and governs their interactions. The properties of energies tell them how to react to similar or different energies. The interactions between similar or different energies define the physical laws of the universe. This is the law of physics.

UR energy is very dense and can be converted to all other energies. To retain the energies of our grand universe, there must also be barriers to contain the energies. In the Ilefos model, this barrier is called a universal domain wall. Read more about this in the chapter "The Universal Domain Wall."

The universal domain wall encircles the grand universe and retains all energies. The beginning of the universe must therefore have been as follows:

1) The presence of the dimensional energy, ER energy
2) The occurrence of large quantities of UR energy
3) A reaction between ER and UR energies forms the universal domain wall that encases the universe.
4) The ER energy then converts UR energy to the other universal energies.
5) These energies give rise to large-scale energy fields in the grand universe.
6) At the point of contact between these fields, the organization of energy systems starts to form.

7) Over time, these organized aggregate into larger formations—similar to those observedn in our universe.
8) These systems will, in time, create black holes, which will grow to their threshold, at which point they cannot control and hold back their energies. The black holes then release their energies in Big Bangs, which replenish the universe with free energy units.
9) The free energy units will be organized into energy systems, such as matter and other systems, eventually creating new big bangs.

As such, our universe began according to the Grand Unified Theory based on the unified force (energy), which was the origin of all universal energies, forming our universe into what we see today. In the Ilefos model, UR energy is the unified energy that formed our universe with ER energy, a dimensional energy.

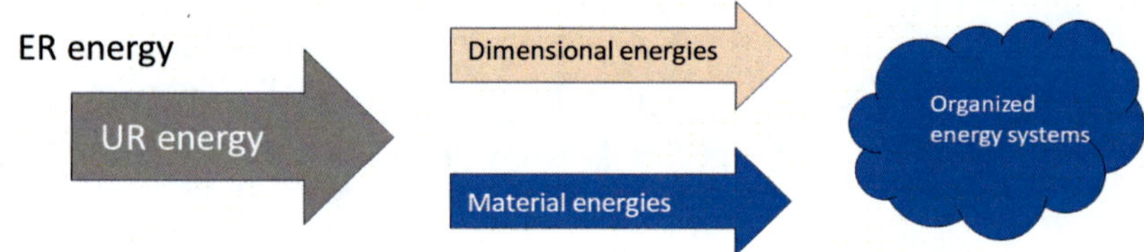

Figure 36. Evolution of the universe. (Credit: Bent Rolf Pettersen).

Chapter 44

Multiverse – Multiple Universes

Our universe is defined by its physical laws, shaped by the properties and interactions of its energies, together with dimensional energies, which carry information.

Our universal energies and their properties characterize our universe.

To preserve and contain its energies, each universe is enclosed by a universal domain wall. This barrier reflects both free energies and organized energy systems. In this way, the universal domain wall retains the energies in the universe, so that it can maintain its energies and energy level. The universal domain wall maintains the universe's internal energy balance.

Figure 37. Multiverses. (Credit Detlev Van Ravenswaay/Science Photo Library).

Our grand universe exists among multiple universes, each with its own energy mixture and specific energies. Each universe may host distinct energies with unique properties. They may also possess different dimensional energies, resulting in unique laws of behavior and organization.

All universes share certain unified energies—primarily dimensional energies that enable information exchange across universes. A creative dimensional energy, along with a specific multiverse energy, enables the existence of multiverses and make it possible to administer them. The special multiverse energy exerts an attractive force on all universal domain walls. This facilitates the coordination and cohesion of universes within in a multiverse structure.

Chapter 45

The Ilefos Model

This book is based on a new model of physics, the Ilefos model. It is an conceptual and qualitative model of energies and their organization. The model is grounded in the properties of energies and their interactions. Each energy may be concentrated and structured into specialized quarks or other organized systems. Quarks may combine into larger energy systems, such as those found in matter and dark matter.

The Ilefos model also explains how organized energy systems grow and behave. Dimensional energies carry information, interact with organized energy systems, and may form specialized dimensional quarks that assist material energy systems. They regulate the behavior of material energy systems. Together, they form the material energy systems such as we know them.

The model describes gravity as an attractive force governed by ilefos dynamics, dark matter, dark energy, and the nature of stars and black holes. In a larger context, it also explains the dynamics of the universe, including how Big Bangs occur. Its explanation of gravity is aligned with quantum mechanics and astrophysics and demonstrating how a single attractive force underlies both atomic bonding and gravitational attraction.

This book demonstrates how the Ilefos model can be used as a universal model may serve as a universal framework for explaining the full spectrum of natural phenomena. It presents is an alternative conceptual, and qualitative universal model grounded in the principles of the Ilefos framework.

Chapter 46

The Quark Periodic System

To explain the properties of atoms in the ilefos model, we must take a look at the quark composition of each type of atom.

The number of plus quarks (ilefos quarks) indicates the potential strength of ilefos tracks (gravitational tracks) in an atom. The strength of ilefos tracks gives us the strength of atomic bonds and gravity.

- One plus quark provides weak ilefos (gravity) tracks, such as we see in gases.
- Two plus (ilefos) quarks potentially give stronger ilefos tracks, such as we see in solid materials.
- Three or more plus quarks may result in very strong ilefos tracks, such as we see in metals.

The strength of ilefos tracks (gravity tracks) is also determined by the energy level of the atom, and the number of other essential quarks. Each atom seems to prioritize the energy it has the greatest capacity to store. Its quark composition therefore also plays a role in the strength of the ilefos tracks.

The number of energy quarks (dielectricity quarks) gives us the potential strength of dielectricity tracks and the dielectricity storage capacity of the atom. Atoms with many plus quarks exhibit high capacity for storing, transporting, and releasing dielectric energy. Atoms with only one energy quark have low energy conductivity and weak dielectricity tracks.

- One energy quark provides weak dielectricity tracks and low dielectricity storage capacity.
- Two energy quarks give potentially moderate dielectricity tracks and dielectricity storage capacity.
- Three or more energy quarks give us potentially very strong dielectricity tracks and dielectricity storage capacity.

The energy level of the atom also determines the strength of the atom's dielectricity tracks. If an atom has a low energy level, it will try to preserve its energy, and may not produce dielectricity tracks. Atoms with many energy quarks seem to reduce the level of other tasks such as producing ilefos tracks. Therefore, the number and the composition of quarks determine how much the atom prioritizes, for example, dielectricity tracks.

Different dimensional quarks are present in all atoms. They provide the potential for communication and also some other tasks. I have included a column in the table below showing "Other dimensional quarks." This tells us if the atom's elements (protons/neutrons) have extra (more than one) special dimensional quarks. Atoms with more than one of these dimensional quarks have extra dimensional energy capacity. This is referred to as "Radioactive matter" due to its extra dimensional energy, which results in extra dimensional energy radiation.

Each element is listed along with its atomic number. This tells us how many protons the element has. Usually, the element has the same number of neutrons, but this can vary in isotopes (variants) of the matter. Hydrogen has no neutrons, except in special isotopes of hydrogen.

Ilefos potential is calculated as the number of neutrons multiplied by the number of plus quarks in the element. This tells us the number of ilefos tracks and their potential energy level. (Each neutron sends out one ilefos track).

The energy potential is listed by the number of atomic elements (protons) multiplied by the number of energy quarks. This gives us the energy (dielectricity) potential of the element and the potential strength of the dielectricity tracks. (Each proton sends out one dielectricity track).

Together, this information will help us explain the properties of each atomic element. I have put these together in an alternative periodic system, the Quark Periodic Table, or the Q periodic table.

The Quark Periodic Table (Qperiodic Table)

Number atom elements*	Name		Plus quarks	Energy quarks	Extra Dim. quarks	Category	Category 2	Ilefos potential	Energy potential
1	Hydrogen	H	0	2	0	Gas	Matter	0,1	2
2	Helium	He	1	2	0	Gas	Matter	2	4
3	Lithium	Li	2	2	0	Light matter	Matter	6	6
4	Beryllium	Be	2	1	0	Light matter	Matter	8	4
5	Boron	B	2	1	0	Light matter	Matter	10	5
6	Carbon	C	2	2	0	Light matter	Matter	12	12
7	Nitrogen	N	1	1	0	Gas	Matter	7	7
8	Oxygen	O	2	3	0	Gas	Matter	16	24
9	Fluorine	F	1	1	0	Gas	Matter	9	9
10	Neon	Ne	1	1	0	Gas	Matter	10	10
11	Sodium	Na	2	1	0	Light matter	Matter	22	11
12	Magnesium	Mg	2	2	0	Light matter	Matter	24	24
13	Aluminum	Al	3	2	0	Light matter	Matter	26	26
14	Silicon	Si	2	2	0	Light matter	Matter	28	28
15	Phosphorus	P	2	3	0	Light matter	Matter	30	45
16	Sulfur	S	2	3	0	Light matter	Matter	32	48
17	Chlorine	Cl	1	3	0	Gas	Matter	17	51
18	Argon	Ar	1	1	0	Gas	Matter	18	18
19	Potassium	K	2	3	0	Light matter	Matter	38	57
20	Calcium	Ca	2	2	0	Light matter	Matter	40	40
21	Scandium	Sc	2	1	0	Light matter	Matter	42	21
22	Titanium	Ti	3	2	0	Metal	Matter	66	44
23	Vanadium	V	3	1	0	Metal	Matter	69	23
24	Chromium	Cr	3	2	0	Metal	Matter	72	48
25	Manganese	Mn	3	2	0	Metal	Matter	75	50
26	Iron	Fe	4	3	0	Metal	Matter	104	78

(Continued)

Number atom elements*	Name		Plus quarks	Energy quarks	Extra Dim. quarks	Category	Category 2	Ilefos potential	Energy potential
27	Cobalt	Co	3	3	0	Metal	Matter	81	81
28	Nickel	Ni	3	1	0	Metal	Matter	84	28
29	Copper	Cu	3	4	0	Metal	Matter	87	116
30	Zinc	Zn	3	1	0	Metal	Matter	90	30
31	Gallium	Ga	3	1	0	Metal	Matter	93	31
32	Germanium	Ge	3	1	0	Metal	Matter	96	32
33	Arsenic	As	3	1	0	Metal	Matter	99	33
34	Selenium	Se	3	1	0	Metal	Matter	102	34
35	Bromine	Br	3	1	0	Metal	Matter	105	35
36	Krypton	Kr	1	1	0	Gas	Matter	36	36
37	Rubidium	Rb	3	1	0	Metal	Matter	111	37
38	Strontium	Sr	3	1	0	Metal	Matter	114	38
39	Yttrium	Y	3	1	0	Metal	Matter	117	39
40	Zirconium	Zr	3	1	0	Metal	Matter	120	40
41	Niobium	Nb	3	1	0	Metal	Matter	123	41
42	Molybdenum	Mo	3	1	0	Metal	Matter	126	42
43	Technetium	Tc	3	1	0	Metal	Matter	129	43
44	Ruthenium	Ru	3	1	0	Metal	Matter	132	44
45	Rhodium	Rh	3	1	0	Metal	Matter	135	45
46	Palladium	Pd	3	1	0	Metal	Matter	138	46
47	Silver	Ag	3	1	0	Metal	Matter	141	47
48	Cadmium	Cd	3	1	0	Metal	Matter	144	48
49	Indium	In	3	1	0	Metal	Matter	147	49
50	Tin	Sn	3	1	0	Metal	Matter	150	50
51	Antimony	Sb	3	1	0	Metal	Matter	153	51
52	Tellurium	Te	3	1	0	Metal	Matter	156	52

Number atom elements*	Name		Plus quarks	Energy quarks	Extra Dim. quarks	Category	Category 2	Ilefos potential	Energy potential
53	Iodine	I	3	1	0	Metal	Matter	159	53
54	Xenon	Xe	1	1	0	Gas	Matter	54	54
55	Cesium	Cs	3	1	0	Metal	Matter	165	55
56	Barium	Ba	2	1	0	Light matter	Matter	112	56
57	Lanthanum	La	2	1	0	Light matter	Matter	114	57
58	Cerium	Ce	2	1	0	Light matter	Matter	116	58
59	Praseodymium	Pr	2	1	0	Light matter	Matter	118	59
60	Neodymium	Nd	3	2	0	Metal	Matter	180	120
61	Promethium	Pm	2	1	0	Light matter	Matter	122	61
62	Samarium	Sm	2	1	0	Light matter	Matter	124	62
63	Europium	Eu	2	1	0	Light matter	Matter	126	63
64	Gadolinium	Gd	2	1	0	Light matter	Matter	128	64
65	Terbium	Tb	2	1	0	Light matter	Matter	130	65
66	Dysprosium	Dy	2	1	0	Light matter	Matter	132	66
67	Holmium	Ho	2	1	0	Light matter	Matter	134	67
68	Erbium	Er	2	1	0	Light matter	Matter	136	68
69	Thulium	Tm	2	1	0	Light matter	Matter	138	69
70	Ytterbium	Yb	2	1	0	Light matter	Matter	140	70
71	Lutetium	Lu	2	1	0	Light matter	Matter	142	71
72	Hafnium	Hf	3	1	0	Metal	Matter	216	72
73	Tantalum	Ta	3	1	0	Metal	Matter	219	73
74	Tungsten	W	3	1	0	Metal	Matter	222	74
75	Rhenium	Re	3	1	0	Metal	Matter	225	75
76	Osmium	Os	3	1	0	Metal	Matter	228	76
77	Iridium	Ir	3	1	0	Metal	Matter	231	77
78	Platinum	Pt	3	1	0	Metal	Matter	234	78
79	Gold	Au	3	1	0	Metal	Matter	237	79
80	Mercury	Hg	3	1	0	Metal	Matter	240	80

(Continued)

Number atom elements*	Name		Plus quarks	Energy quarks	Extra Dim. quarks	Category	Category 2	Ilefos potential	Energy potential
81	Thallium	Tl	3	1	0	Metal	Matter	243	81
82	Lead	Pb	3	1	0	Metal	Matter	246	82
83	Bismuth	Bi	3	1	0	Metal	Matter	249	83
84	Polonium	Po	3	1	0	Metal	Matter	252	84
85	Astatine	At	3	1	0	Metal	Matter	255	85
86	Radon	Rn	1	2	0	Gas	Matter	86	172
87	Francium	Fr	2	1	0	Light matter	Matter	174	87
88	Radium	Ra	4	3	1	Radioactive matter	Light dark matter	352	264
89	Actinium	Ac	3	2	1	Radioactive matter	Light dark matter	267	178
90	Thorium	Th	4	3	1	Radioactive matter	Light dark matter	360	270
91	Protactinium	Pa	3	2	1	Radioactive matter	Light dark matter	273	182
92	Uranium	U	4	4	2	Radioactive matter	Light dark matter	368	368
93	Neptunium	Np	4	3	1	Radioactive matter	Light dark matter	372	279
94	Plutonium	Pu	4	3	2	Radioactive matter	Light dark matter	376	282
95	Americium	Am	4	3	2	Radioactive matter	Light dark matter	380	285
96	Curium	Cm	4	3	2	Radioactive matter	Light dark matter	384	288
97	Berkelium	Bk	4	3	2	Radioactive matter	Light dark matter	388	291
98	Californium	Cf	4	3	2	Radioactive matter	Light dark matter	392	294
99	Einsteinium	Es	4	3	2	Radioactive matter	Light dark matter	396	297
100	Fermium	Fm	4	3	2	Radioactive matter	Light dark matter	400	300
101	Mendelevium	Md	4	3	2	Radioactive matter	Light dark matter	404	303
102	Nobelium	No	4	3	2	Radioactive matter	Light dark matter	408	306
103	Lawrencium	Lr	4	3	2	Radioactive matter	Light dark matter	412	309
104	Rutherfordium	Rf	4	3	2	Radioactive matter	Light dark matter	416	312
105	Dubnium	Db	4	3	2	Radioactive matter	Light dark matter	420	315

Number atom elements*	Name		Plus quarks	Energy quarks	Extra Dim. quarks	Category	Category 2	Ilefos potential	Energy potential
106	Seaborgium	Sg	4	3	2	Radioactive matter	Light dark matter	424	318
107	Bohrium	Bh	4	3	2	Radioactive matter	Light dark matter	428	321
108	Hassium	Hs	4	3	2	Radioactive matter	Light dark matter	432	324
109	Meitnerium	Mt	4	3	2	Radioactive matter	Light dark matter	436	327

* Number of atom element: neutron, proton.

Chapter 47

Articles

The following chapters contain a selection of articles elaborating on specific topics within the conceptual framework of the ilefos model.

Chapter 48

Paradigm Shifts and Research

(Article in "Naturen" no. 6 2023, by Bent Rolf Pettersen).

Bent Rolf Pettersen (b. 1967) is a physicist at Isource. He has conducted extensive work on basic physics, quantum physics, and astrophysics. He will soon publish a book on these issues, related to atomic bonds and gravity.

Over many decades, we have tried to find answers to several of nature's fundamental properties, without success. What is gravity? What is dark energy and dark matter? Why is the universe expanding at an accelerating rate? Why are we unable to reconcile the theories surrounding nuclear physics with astrophysics? Are we bound to scientific frameworks that do not enable answers to these questions? Professor Thomas Kuhn referred to these issues as research paradigms. Are we working within a research paradigm that cannot provide solutions? Do we need a paradigm shift? This chronicle deals with these questions related to current theories in physics.

If you have a problem you cannot solve, you probably need to start over and approach the problem from a different angle. What seems unsolvable often yields to new solutions when the angle of approach and underlying assumptions are changed. This is a philosophical approach to challenges we have trouble solving.

Gravitation, dark energy, dark matter, and the accelerating expansion of the universe; these are some of nature's greatest mysteries that scientists have not been able to explain. We can describe and calculate the phenomena, but why are we unable to explain them?

> "It is recognized that theories are in use only because there are no better alternatives, not because they represent the final answer. You realize that it is not possible to come up with new solutions, and thus new knowledge, but that it is necessary."

Professor Per Arne (Bjorkum, P.A. "Annerledestenkerne," Universitetsforlaget 2016).

Today's scientific theories about how the atom is put together are based on a more than 120-year-old model about the construction of the atom. Theories about energies and gravity are also often based on older models.

Natural science is mostly presented in mathematical terms and formulas, enabling calculations without necessarily considering cause and effect. If you want new knowledge about nature, you must think about how things are connected and cause and effect.

Even if one sees that observations do not comply with the theory, one will not abandon this theory until there is an alternative theory.

This paradox can represent an obstacle to innovation in research: Most researchers prefer scientific findings that support what they already know, not findings that challenge the foundations of their careers and professional identity.

When a revolutionary innovation is launched, it can put the entire research community to the test, both intellectually and emotionally. When irregular thoughts arrive, they can bring about large and dramatic changes in theories about nature. Such dramatic changes can be called paradigm shifts (Thomas Kuhn, T.S., 1962. "On the Structure of Scientific Revolutions." Chicago: The University of Chicago Press).

A paradigm shift means you must build new theories almost from scratch. You have to start with new definitions and concepts and establish new connections that turn established beliefs and theories upside down. Established scientists' views of the world can then feel threatened. Most scientists are guided by their theories, and they have strong feelings about them. The theories often merge with the scientists, becoming part of their identity.

If a problem remains unsolved after decades of intense research, it is reasonable to assume that the research community is not working with the right pieces of the puzzle. They are probably working within a paradigm that does not enable a solution.

If you are unaware of this possibility, you remain in the paradigm without the possibility of solving the task you have set out to solve. New observations can start paradigmatic debates, but the absence of solutions should also start such debates.

The reason why one gets stuck in a paradigm is that it is common to perceive theories that are recognized and practiced within an established paradigm as proven. Scientists are often not trained to evaluate the paradigm itself.

When a problem has existed for some time without anyone being able to solve it, it is usually the methodological approach that needs to change. The issue must be tackled in a new way.

One can solve some, but not all, problems if one is faithful to a paradigm, that is to say, to the established methodology. Yet often, a greater and greater discrepancy will accumulate between theory and observations.

Today, we have many applicable theories depending on what we want to explain. In physics, there are different theories that deal with particle physics and astrophysics. The standard model, string theory, quantum mechanics, and relativity are some of them. Each theory is often adapted to its specific area and must be changed if it is to be applied to other areas. There is no theory that can be used in all areas. Correct physics must work in all areas.

Physicist Stephen Hawking spent large parts of his life trying to unify the theories into a universal theory with application in all areas. For such a theory to arise, we must have a paradigm shift in physics. We must dare to think outside established theories and models.

Only then will we achieve a paradigm shift that leads to a better understanding of nature.

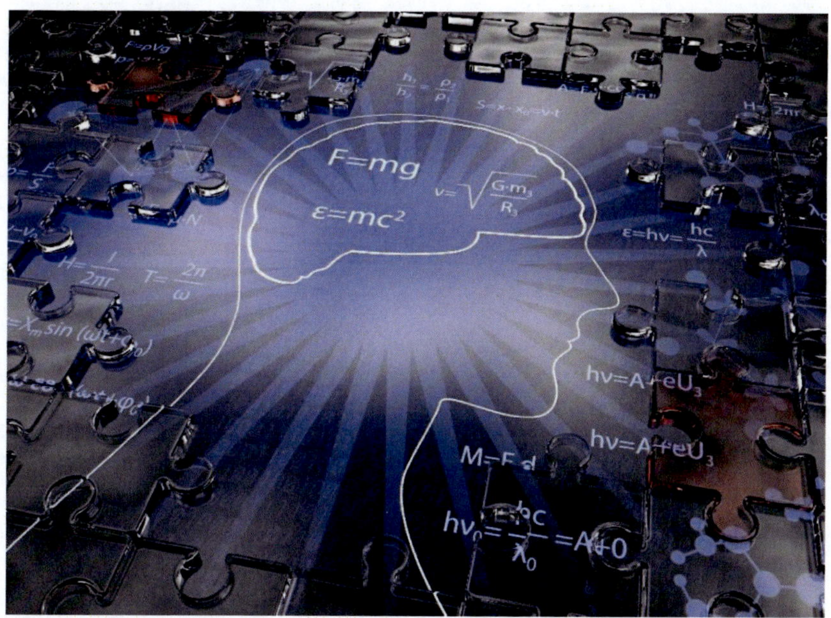

Figure 38. Brain (iStock).

Chapter 49

The Attractive Force: One Force Creates Atomic Bonds and Gravity

Introduction

Can we find an explanation of the attractive force between atoms that works with both quantum mechanics and astrophysics?

An attractive force pulls atoms towards each other. This force is very strong close to the atomic core, where it can form strong bonds, like those we see in metals. Further from the nucleus, an attractive force seems to form a weaker bond between atoms, gravity. Universal phenomena like stars and black holes also emit an attractive force that works at long distances from the source. Could a single attractive force be responsible for all these bonds?

We have theoretical models explaining the nuclear bindings and gravity in the universe, but we have trouble combining these theories into one model that works with both quantum mechanics and astrophysics. String theory is our best attempt to explain atomic bonds and gravity in one model.

Our present theories apply hypothetical elements in order to work. Yet, after decades of intense research, we are still unable to explain the basics of nature, such as universal gravity, dark energy, dark matter, and why the universe expands at an accelerating speed higher than the speed of light. Are we bound within a scientific framework that does not enable answers?

Professor Thomas Kuhn [1] referred to these issues as research paradigms. Are we working within a research paradigm that does not allow a solution? Is the lack of answers an indication that we need a paradigm shift?

If you have a problem you cannot solve, you must probably start over and attack the problem from a different angle. What seems unsolvable often has a solution if we change the angle of attack and the rules.

In this article, I do not, therefore, bind myself to present theories within the different fields but focus on what properties the attractive force must have, and how we can create an explanatory model that explains the attractive force in both quantum mechanics and astrophysics.

This model must explain the attractive force, why it decreases in strength with distance from its source, and why different sources of the attractive force vary in strength.

The Attractive Force

In order to explain this force, we must look at what kind of energy could generate a force. Could the properties of different energies, and how they influence and interact with each other, create forces?

In this model, I focus on an energy with attractive properties. This attractive energy is referred to as ilefos, the attractive energy, and the model the Ilefos model.

If pulses of ilefos energy are attracted to each other, the attractive force between the ilefos energy units will make them form a track when released from the source. In the following, I endeavor to explain this in relation to atoms, but I will also show how this also works in the context of astrophysics.

When ilefos energy units are emitted from the nucleus, the mutual attractive force between the units makes them form a track of ilefos energy units, an ilefos track. This is a gravity track. In the model, each neutron emits an ilefos track. Hydrogen does not have a neutron, so the proton has to emit the gravity track, but this is much weaker than the tracks of neutrons.

An ilefos track has attractive properties towards external ilefos tracks stemming from other atoms. This attraction will draw them toward each other and make them connect. The strength of the ilefos track determines the strength of the connection. If two strong ilefos tracks connect, we can have a strong connection, an atomic bond. If two weaker tracks meet, we have a weaker connection, gravity. The strength of the bond is determined by the weakest ilefos track.

We must then explain why different atoms have different strengths in their ilefos tracks. We must assume that different atoms have different quark compositions, which give the atom its properties. The atom must have one or more quarks responsible for storing and handling ilefos energy. I call this quark a plus quark. This could be equivalent to the Higgs boson. The strength of the ilefos is determined by the number of plus quarks (ilefos quarks) in the neutron. If a neutron only has one plus quark, it releases a weak ilefos track, such as those seen in gases. If the neutron has two plus quarks, it creates a stronger ilefos track, as seen in solid matter. If a neutron has three or more plus quarks, it releases a strong ilefos track, as seen in metals.

The general energy level of the atom also determines the strength of the ilefos track. At very low temperatures, the atom prioritizes other tasks and energies, like the strong nuclear force. The atom will then emit weaker ilefos tracks, which in turn will create weaker atomic bonds.

The Decrease of the Attractive Force with the Length of a Gravity Track

To explain why the attractive force is very strong close to the source where it forms atomic bonds, and then weakens with distance, we must look at different energies and their properties, and how they interact.

Ilefos energy units have an attractive force toward external ilefos energy units.

The strong nuclear force is an energy that holds the atomic core together in a fixed and organized manner. Different energies are concentrated and organized in different quarks. Different quarks are organized into groups of quarks, protons, and neutrons. These are, in turn, organized in an atomic structure. The strong nuclear force keeps the quarks and the groups of quarks together in a solid structure. We can therefore also assume that the strong nuclear force has an attractive force toward ilefos.

If ilefos is attracted to the strong nuclear force, the ilefos energy units will experience resistance when they are emitted from the neutron. The drag from the strong nuclear force will reduce the units' speed when influenced by the strong nuclear force, causing a strong concentration of ilefos energy pulses in the ilefos track close to the atom.

With distance from the source, the influence of the strong nuclear force will decrease. A weaker influence results in reduced drag, and the ilefos units increase in speed in the ilefos track. When the ilefos energy units increase in speed, we will have a greater distance between the ilefos energy units in the ilefos track. The attractive force of the ilefos will then be reduced. The energy units will continue to accelerate until they achieve their natural energy speed. This will result in a reduced strength of the gravity track, much like Isaac Newton's law of universal gravitation [2].

$$F = G \frac{m_1 m_2}{r^2}$$

F is the gravitational force acting between two objects, m1 and m2 are the masses of the objects, r is the distance between the centers of their masses, and G is the gravitational constant.

This law may apply to the decrease of strength in an ilefos track due to the accelerating speed of the ilefos energy units. However, we also need to add an element to calculate the influence of the strong nuclear force close to the atom. This influence dramatically reduces the speed close to the atom, thus making the ilefos track very strong, which explains the creation of atomic bonds.

The first element in the gravitational formula will explain how the influence of the strong nuclear force creates a drag that reduces the speed of the ilefos energy units, thus helping to keep the gravity track strong close to the source.

The following is a proposed formula for the attractive force (ilefos) close to the nucleus:

$IP = (Sn \times IP_1)/r^4$
IP= Ilefos power, the strength of the ilefos track at a given point
Sn= Strong nuclear force of the atom
IP_1= Ilefos power at the track's starting point
r= the distance from the source (in pm, picometres)

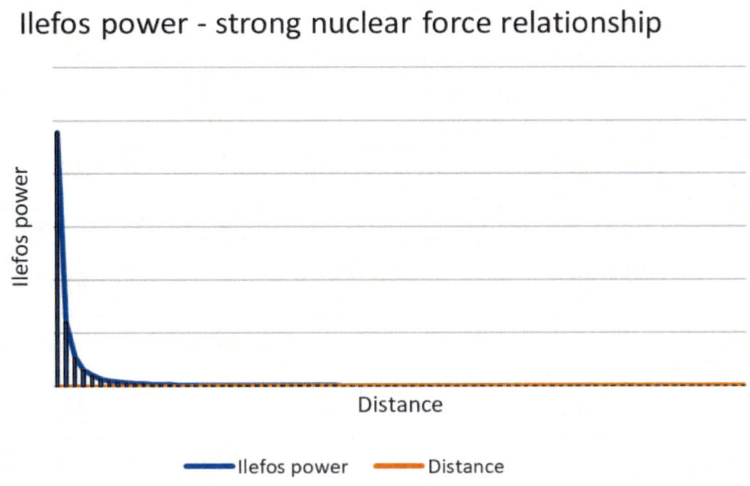

Ilefos power - strong nuclear force relationship
Ilefos power
Distance
Ilefos power
Distance

Figure 39. The power of an ilefos track in relation to the strong nuclear force. (Bent Rolf Pettersen).

For this formula to work, the distance r must be incredibly small.

The distance, r, cannot be given in meters. The new r is here is picometres, pm, or 1×10^{-12} m. An atom's size, including the "electronic cloud," is between 62 and 520 pm in diameter. The size of a nucleus is between 0.002 and 0.012 pm. r in this formula is therefore measured in picometres/pm.

This first formula gives us the ilefos power, or strength of the ilefos track, close to the atom, where the strong nuclear force affects the movement of the ilefos energy units in the track.

This is a provisional formula indicating the effect of the strong nuclear force in relation to the attractive force.

With distance from the atom, the influence of the strong nuclear force diminishes and eventually disappears. As such, this first part of the formula will lose its value, and Isaac Newton's formula of universal gravitation takes over. The law of universal gravitation replaces the first part, the strong nuclear force influence part, when this value is greater than in the first formula.

$$F = G\frac{m_1 m_2}{r^2}$$

$F=G (m_1 \times m_2)/r^2$

F= The force acting between two masses

G= Gravitational constant

m= Mass of the objects

r= the distance between the objects (m)

However, we must revise this formula in order to work with ilefos power in an ilefos track.

M_1 and M_2 are replaced by the atomic weight (u).

The new revised formula for ilefos power without the strong nuclear force influence is then:

$IP_2 = G (u^2)/r^2$

IP_2= Ilefos power in the ilefos track when not influenced by the strong nuclear force

G = Gravitational constant

u = the atomic mass of the atom

r = distance between the atom and the measurement point

$IP = (Sn \times IP_1)/r^5 \geq G\,(u^2)/r^2$

The first formula calculates the ilefos power close to the nucleus, where the track is influenced by the strong nuclear force. The second formula calculates the ilefos energy in the ilefos track when the track is not influenced by the strong nuclear force. This is therefore a conditional formula.

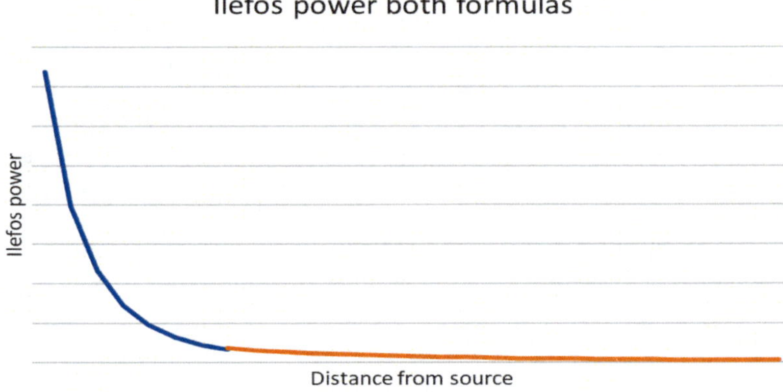

Figure 40. Ilefos power from an atom, both formulas. (Bent Rolf Pettersen).

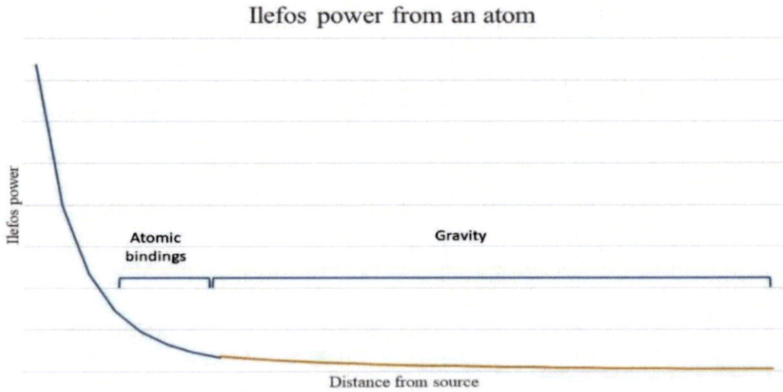

Figure 41. Ilefos power in an ilefos track emitted from an atom. In the first part, the influence of the strong nuclear power keeps the gravity track dense. With distance, this influence diminishes and the ilefos energy units can accelerate without this drag. (Bent Rolf Pettersen).

Here, the blue line shows ilefos power influenced by the strong nuclear force, and the orange line shows ilefos power without the influence of the strong nuclear force. A strong ilefos track can form an atomic bond if it connects to a strong external ilefos track. When it becomes weaker, it can form a weaker connection to an external track, in the form of gravity.

When the influence of the strong nuclear force diminishes, the ilefos power from the first formula quickly reduces ilefos power to approximately zero. Then, the second formula takes over, where the reduction of the ilefos power is reduced by the square of the distance to the source (r^2).

The attraction between two masses will be the strength of the ilefos track at the point where the ilefos tracks from the different masses connect. With larger objects, this means the sum of the strength of multiple ilefos tracks. Here, Isaac Newton's law of universal gravitation [2] is a good approximation.

We should also add a section related to the drag of dark energy. If dark energy is composed of the same energies as we find in matter, it also contains energy elements of the attractive force, ilefos.

In this model, dark energy consists of "free" energy units that are not concentrated and organized in quarks, particles, or matter. According to the Lambda-CDM model [3], 68% of the visible universe is dark energy (Λ). We can therefore assume that free ilefos energy units in dark energy affect the ilefos energy units in an ilefos track (gravity track). Ilefos units in dark energy will then most likely create a weak drag towards the speed of the ilefos energy units in an ilefos track.

In order to have a correct G_1 (modified gravity constant), we should also include a dark energy drag in the formula. This drag is most likely very weak, and I do not, therefore, include it at this point.

Ilefos Gravity Tracks from an Atom

The ilefos gravitational formula shows how the ilefos track is strongly affected by the strong nuclear force in the perimeter around the nucleus. This will make the ilefos track bend and circle outwards. When the attraction of the strong nuclear force weakens with distance from the nucleus, the ilefos energy units in the ilefos track will accelerate. A centrifugal effect will occur, and with the acceleration of the ilefos units in the track, the track will slowly straighten. When the ilefos track is no longer affected by the strong nuclear force, the track will eventually level out to become completely straight.

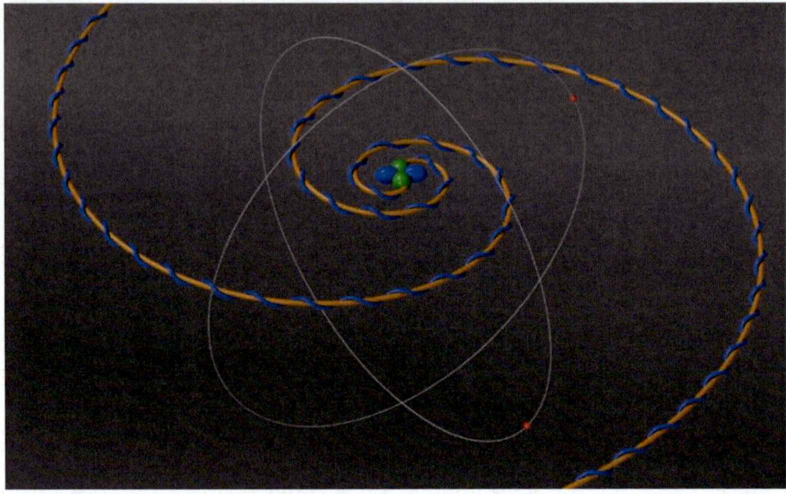

Figure 42. Ilefos track extending from an atom. The attraction from the strong nuclear force produces a Fibonacci-like track, which straightens up with distance from the nucleus. (Credit: Stian Kongsvik).

We have now explained how a gravity track from an atom (ilefos track) can be very strong close to its source, where it can form strong connections to external tracks, atomic bonds. With greater distance from the source, the ilefos track will weaken and create weaker connections, gravity.

The same formula also applies to stars. We can then assume that stars dissolve organized energy systems (matter) and release the surplus energy in strong energy tracks. Stars then also release ilefos, which forms ilefos tracks. These are very strong gravity tracks with a far reach.

Black holes also have a similar dynamic and release ilefos units that form ilefos tracks. Here, however, the internal attraction from the black hole's core is extremely strong. This strong gravity does not allow other energies that are affected by gravity to be released. Therefore, black holes do not release "heat," except through Hawking radiation in areas with weaker gravity.

The Recycling of Energies through Atoms

With distance from the source, an ilefos track (gravity track) becomes very weak. The acceleration of the ilefos energy units results in a greater distance between each unit in the track. Eventually, the distance becomes too long for the mutual attractive force of the ilefos energy units to keep them together in a track. The ilefos track will then dissolve into free ilefos energy units, which are not connected to a source.

Dark energy is free energy units. With the release of ilefos and the dissolving of an ilefos track, we feed the environment with dark energy. Ilefos must therefore make up a significant proportion of dark energy.

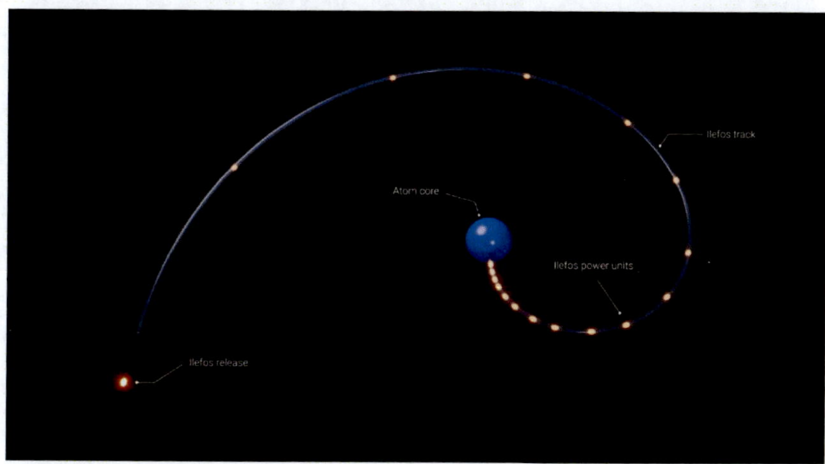

Figure 43. Ilefos energy units in an ilefos track extending from an atom. The pulses of ilefos energy units accelerate with distance from their source. (Credit: Stig Boe/Mohammed Ismaiel).

Atoms also constantly release energy to perform other tasks, such as generating a repulsive force and dielectricity (heat). All of these actions indicate that atoms release energy. To be in energy balance, atoms must receive the same amount of energy.

The Lambda-CDM model calculates the universe to be 68% dark energy. We must assume that dark energy is everywhere, including in the vicinity of atoms. This indicates that there are immense amounts of "free" energy units, which the atoms may harvest and absorb. The atoms are, as such, constantly fed with energy through dark energy. This indicates an energy balance in atoms and shows how atoms can recycle energy units to achieve energy balance.

Atoms continuously expend energy to perform their functions. The energy they receive may not always be the energy they need, and they must therefore convert it. This is called Atomic Phase Displacement [4].

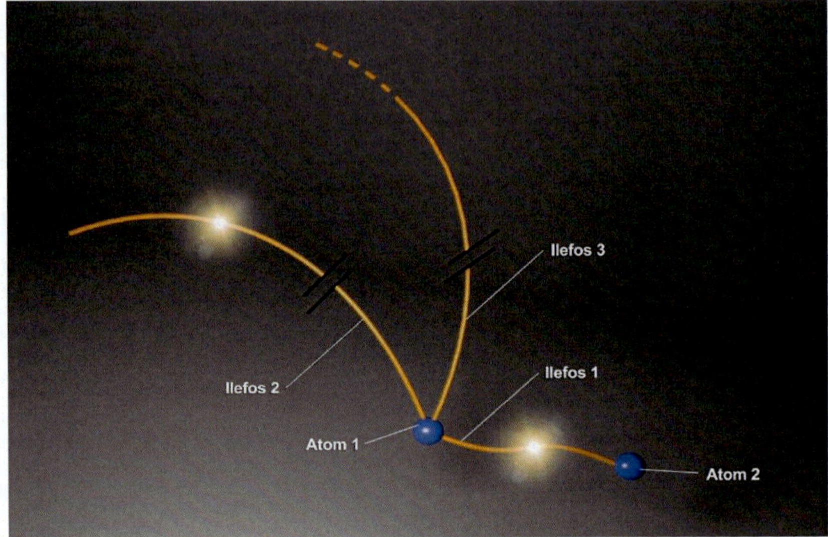

Figure 44. Three types of ilefos tracks extending from an atom. Ilefos 1 forms a strong atomic bond close to the atom. Ilefos 2 forms a weaker bond further out with an external ilefos track, gravity. Ilefos 3 does not connect with an external ilefos track, but becomes very weak, dissolves, and performs an ilefos release. (Credit: Stian Kongsvik).

Energy and Forces

Different energies have different properties. These properties determine how the energy reacts to similar or different energies.

A force is a reaction between energies. Energies create forces based on their properties and interaction with each other.

Conclusion

When we introduce energy units that have a strong attraction force toward each other (ilefos) and the strong nuclear force, we can explain the nature of gravity tracks in a manner that can be applied in both quantum mechanics and astrophysics. The properties of this energy may be attractive forces that work in all fields of the universe.

The ilefos model explains how the ilefos energy units attract each other and form a gravity track when released from the atom. The ilefos energy units experience a strong drag close to the nucleus due to the strong nuclear force. This attraction results in a dense, strong track close to the nucleus. When this strong track connects with an external strong ilefos track, an atomic bond is formed. With distance from the source, the strength of the track becomes weaker. When the weaker track connects with an external track, it forms a weaker connection, gravity.

As such, the ilefos model appears to explain how an ilefos track (gravity track) can generate both strong atomic bonds, like those seen in metals, and gravity. The model would appear to be applicable within both the quantum field and astrophysics.

References

[1] Kuhn. T. S., "The Structure of Scientific Revolutions" (1962).

[2] Isaac Newton's Principia, the mathematical principles of natural philosophy (1687).

[3] Michael S. TURNER, Lambda-CDM model," ΛCDM, Much more than we expected, but now less than what we want," Foundations of Physics, 2018.

[4] Bent Rolf Pettersen, "Atomic Phase Displacement – conversions of energies in atoms," ResearchGate 2023, http://dx.doi.org/10.13140/RG.2.2.11943.32166.

While the simplified ilefos force model captures the fundamental dynamics of attraction, a more refined formulation can account for the differentiated roles of quark types. This allows for improved accuracy in describing near-field binding forces at the atomic and molecular scale.

Refined Track Force Calculation (Ilefos 2.0)

In the ilefos model, the attractive force between atomic structures—whether functioning as chemical bonds or gravity—can be modeled through resonance tracks formed between organized quark structures. These tracks are primarily driven by the number of *plus quarks*, which represent the gravitational or ilefos-attracting component of matter.

However, *energy quarks*, which regulate dielectricity and energetic flux, may contribute to the strength and stability of ilefos tracks in near-field conditions. They do not initiate attraction in the same way as plus quarks, but may enhance or stabilize an existing track by increasing local coherence.

To represent this asymmetry, we define a **quark energy factor** Q as:

$Q=p+\epsilon\cdot e$

Where:

- p = number of plus quarks
- e = number of energy quarks
- ϵ = weighting coefficient for energy quarks (typically between 0.3 and 0.5)

Near-Field Ilefos Force Equation

The potential force between two structures connected via an ilefos track in the near field is then:

IPnear = $Sn \cdot Q \,/\, r4$

Where:

IPnear = ilefos potential in the near field
Sn = near-field nuclear strength factor (related to local YT resonance)
r = distance between structures (typically measured in picometers (pm) or nanometers (nm), depending on scale)

This model preserves the steep attenuation of force with distance, but offers better scaling to resonance behavior observed in molecular interactions.

Far-Field Gravity Equation

At larger distances, ilefos tracks become sparse and weak. In this regime, gravity behaves according to classical field mechanics. The simplified far-field model is:

$IPfar = G \cdot u2 / r2$

Where:

u = ilefos-active quark units in each object
G = proportional constant for ilefos gravity
r = distance between objects

A characteristic transition radius rc (~0.3 nm) may separate the near-field quantum domain from the far-field gravitational regime.

Chapter 50

$E=Mc^2$ versus E=qm

$E=mc^2$ is Albert Einstein's famous mass-energy equivalence formula from 1905. The formula gives us the relationship between mass and energy. The energy (E) of a particle equals the product of mass (M) times the speed of light squared (c^2). The formula tells us that a small amount of mass corresponds to very strong energy.

However, is it right to calculate the energy in matter using the speed of light (c^2)?

E=M tells us that energy and matter are interchangeable. Matter is energy that is organized and concentrated. Matter can store immense amounts of energy. In the mass-energy equivalence formula, c is the speed of light. Light travels in a vacuum approximately 300,000 kilometers per second/186,000 miles per second. C^2 equals 9×10^{16} m^2/s^2 (or approximately 90,000,000,000 kilometers per second/34,596,000,000 miles per second). This immensely high number indicates an immensely strong energy stored in matter. To calculate the kinetic energy of a mass (m) in movement (v=velocity), we often use the formula $\frac{1}{2} m v^2$.

However, kinetic energy can tell us something about how matter can release energy in contact with external matter.

If matter, in the form of atoms, collides with external atoms, both atoms will try to preserve their structure. When two atoms collide, they increase the repulsive force in their uni particles, which orbit the nuclei. The uni particles orbiting the nucleus form a repulsive shield that protects the nucleus.

Increased repulsive energy (uni energy) that is released is friction. If the increased repulsive energy of the uni particles is not enough to prevent a collision between the nuclei, the atoms also will release energy in fission or atom dissolution. If this happens, atomic elements, like neutrons, are knocked away from the nucleus. These elements will dissolve within a short time and release the energy from their quarks. If the atom has enough energy, it will replace the missing elements. If the collision is too strong, the entire atomic structure can be destroyed. The atomic system will then be unable to replace the missing atomic elements and keep the remaining elements together. All the energy from the atomic structure will be released in an atom dissolution.

In friction/collision, atoms release energy, which the atoms will restore over time.

When two objects with equal masses collide, both masses release equal amounts of energy. If one object does not move and we transfer movement (kinetic energy) to the other object, both objects release the same amount of energy. We could say that kinetic energy is the relative movement between the objects, but is this correct? Do matter and atoms gain energy with speed?

Atoms do not gain energy with speed; rather, they lose energy through interactions and collisions with external atoms. The speed tells us something about the potential to lose energy in contact with external atoms, but the energy in the atom, in the atom's quarks, does not change due to speed.

Back to the speed of light, c^2. If the energy in matter does not increase with speed, and kinetic energy only tells us the potential for the release of energy, should c be in the formula?

I propose a new formula where energy in matter is not dependent on the speed of light.

Energy in matter is a constant and is not affected by speed. The energy in matter is organized and concentrated in quarks, which serve as energetic carriers and structural nodes. The energy in quarks can be altered by energies in the environment, but not due to speed.

The new proposed formula is:

$$E=qm$$

E=qm shows us that energy in matter is constant, independent of speed (c^2). Energy = the mass of the quarks (qm) is an approximation in which the energy is measured by the atom's mass. Mass reflects the energy density the attractive force of an atom, primarily governed by ilefos interactions. This force is responsible for atomic bonds and gravity.

On this basis, we only take into account energies that are affected by gravity, which is ilefos (gravitation), the strong nuclear force, and dielectricity (electricity). Quarks can also store energies not affected by gravity, such as the repulsive force and dimensional energies. These energies are not taken into account in this formula, but it nonetheless provides a good approximation.

E=qm + energies not affected by gravity appears to be a correct formula. Other energy (Oe) efers to quark-bound energies that are not governed by gravitational or attractive dynamics, such as repulsive and dimensional forces. The new formula is then:

$$E = qm + oe\ (T)$$

This formula is a calculation of the energies in matter related to the energies stored in the atom's quarks.

In the quark periodic table, we list some quarks and their presence in atoms. The energy level in these quarks can also vary, and temperature can provide an indication of this. Therefore, temperature (T) should also be taken into account in the formula.

Figure 45. A nuclear bomb shows some of the immense energy of atoms. (iStock).

Bent Rolf Pettersen, 2024.

Chapter 51

Fusion

In fusion, two atoms merge to form one larger atom.

For a fusion to take place, two things are required:

1) Extremely high energy
2) Matter moving in the same direction

1. Extremely High Energy

Fusion requires a lot of energy. A lot of energy means an atom's uni particles orbit in wide orbits. Uni particles exert a repulsive force on other uni particles. This ensures that atomic nuclei do not collide, and it t stabilizes ilefos orbits (gravity tracks).

When an atom receives a lot of energy, its uni particle orbits become wider and powerful, creating greater distances between the atomic nuclei. At the same time, the atom acquires very strong ilefos tracks (attractive force/gravity tracks). Ilefos tracks pull atoms towards each other. We thus get a strong attraction between the atoms simultaneously as the powerful uni particles push the atomic nuclei away from each other.

Eventually, the uni particle trajectories of different atoms can become so wide that they cross each other's orbits. If this happens, the powerful ilefos tracks can pull atomic nuclei close to each other without the uni particles being able to prevent this.

This causes the nuclei of the two atoms to collide. This can take the form of a powerful or a soft nuclear collision.

A strong collision between atomic nuclei means that the nuclei are completely or partially broken into pieces in an atomic dissolution. Vast amounts of energy are then released from the quarks of the nuclear elements, which then dissolve.

A soft collision does not have enough energy to break apart nuclear elements. This is more of a softer touch between atomic nuclei. In a soft collision between two atomic nuclei, the nuclei are pulled towards each other and fuse into a larger atom, consisting of nuclear elements from both atoms. If a carbon atom (6 neutrons and protons) fuses with a nitrogen atom (7 neutrons and protons), these become an atom with 13 neutrons and protons, i.e., aluminum. The uni particles of the two atoms are retained by the new atom. The new atom then gets 13 uni particles controlled by the new atom's neutrons.

Each atom has an atomic quark that controls its operating system. This also contains the atom's DNA. (DNA is a misleading word since it is linked to living organic matter, but the term is used since the tasks are similar.) Atomic quarks thus have the blueprint for the atom's structure. An atom cannot have two atomic quarks. The atomic quark of the smallest atom will therefore be dissolved. If two identical atoms fuse, the atoms agree on which atomic quark should dissolve.

The new atomic quark has the blueprint for the new atom. The composition of quarks in the nuclear elements (protons/neutrons) and uni particles of the new atom is adapted according to the new atom's DNA. Nuclear elements can thus have more or fewer nuclear quarks, in the form of plus quarks and energy quarks. In addition, several special quarks are changed to adapt to the new atom's DNA and operating system.

2. Matter Moving in the Same Direction

When atomic nuclei collide, this usually results in a violent collision and nuclear disintegration, such as in stars and black holes. When the outcome is not nuclear disintegration, a soft collision must have occurred.

For a soft collision between atomic nuclei to occur, the atoms must move in the same direction. Then, the kinetic energy of both atoms will be applied in the same direction, and the collision will be less powerful, i.e., a soft collision.

In addition, the repulsive force/shield around the atoms must be dismantled. This happens when the uni particles' orbits cross each other. The repulsive force of the uniparticles will then no longer be able to protect the nuclei.

In a soft collision, atomic nuclei are not broken into pieces, but fuse.

In a fusion, quarks or nuclear elements are not dissolved. The fused atoms then retain their energy. A normal fusion process does not, therefore, produce energy.

Figure 46. Illustration of a hard and soft collision between atomic nuclei (Bent Rolf Pettersen).

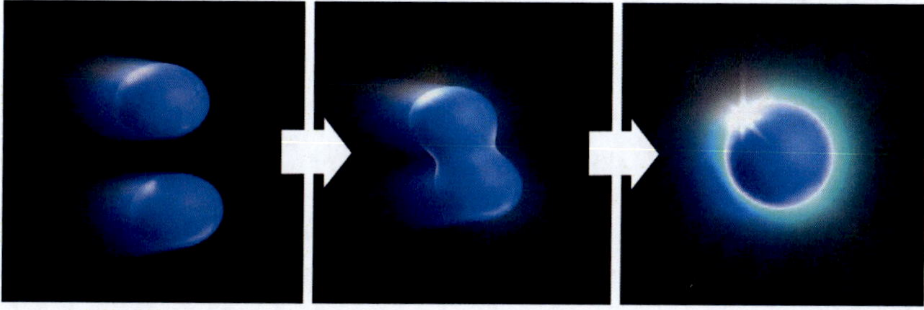

Figure 47. Illustration of a soft collision between atomic nuclei – fusion. (Credit: Ola Tandstad).

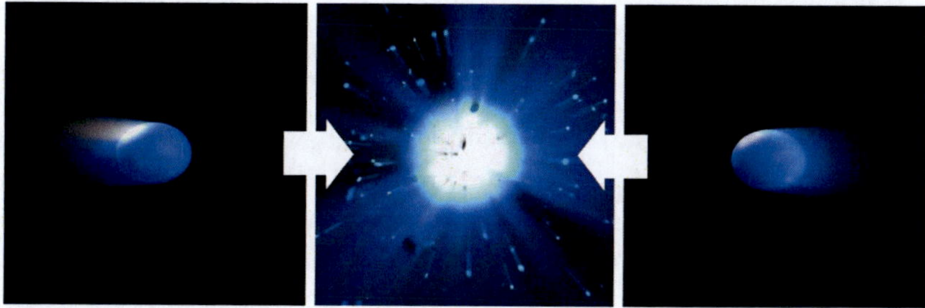

Figure 48. Illustration of a hard collision between atomic nuclei - atomic dissolution. (Credit: Ola Tandstad).

In the case of a hard nuclear collision, we get complete or partial nuclear dissolution. The atomic nuclear elements are then broken, and quarks dissolve and release their energy. A nuclear dissolution releases a vast amount of energy.

A normal fusion requires a lot of energy and is a soft collision. This event does not normally release energy.

Fusion requires powerful energy and atoms moving in the same direction. We find these conditions in several places in the universe.

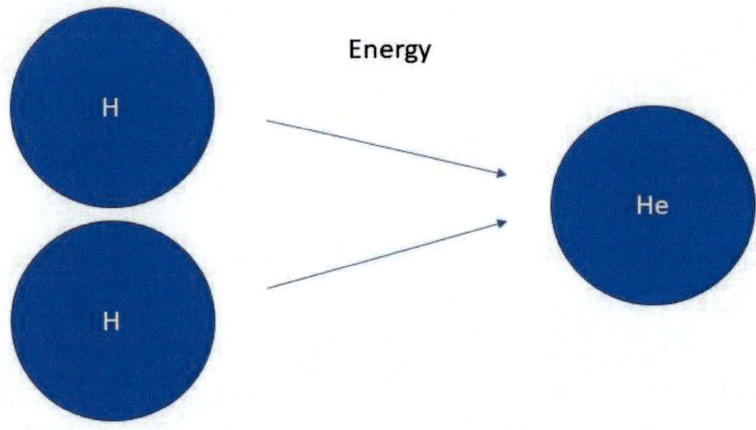

Figure 49. Illustration of fusion of two hydrogen atoms to helium. (Credit Bent R. Pettersen).

In the Big Bang, a lot of matter was moving in the same direction with an extremely high amount of energy. A large portion of the matter in the universe was created right after the Big Bang, when all matter moved outwards from the center. Matter was created by the merging of quarks and the fusion of atoms.

When massive stars at the end of their lives explode in a supernova, a lot of energy is released. Energy, matter, and quarks are thrown from the center, resulting in a substantial amount of matter moving in the same direction with a lot of energy in the surroundings. The conditions for fusion are present, and matter is created. Many of the heavier atoms are created in connection with supernova explosions.

The conditions for fusion are also present in several other galactic events. In Hawking radiation from black holes and in energy beams from quasars, a lot of energy is present. If matter also moves in the same direction in such energy beams, fusion can also occur, although this is rare.

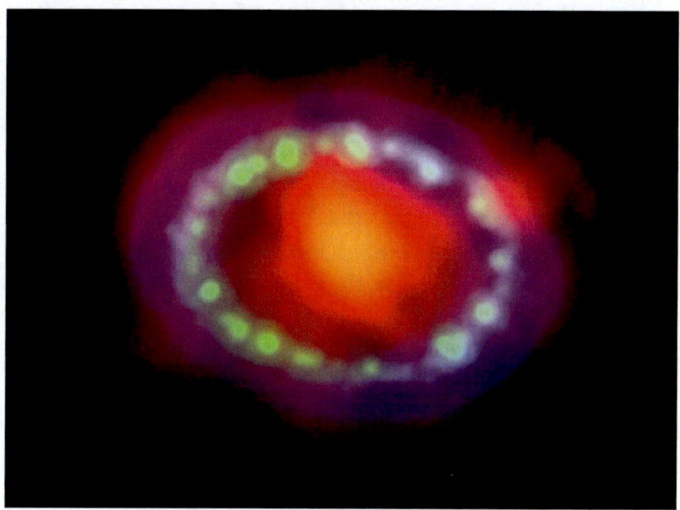

Figure 50. Supernova 1987A (NASA-ESA, A. Angelich).

Atomic Interactions

Four types of interactions between atoms

	Atom dissolution	Fusion	Fission 1	Fission 2
Uni particle orbits	Extremely wide	Extremely wide	Very wide	Wide
Collision	Extreme powerful	Weak	Very powerful	Moderate
Crushed neutrons	1- … (all)	0	1-2	0
Energy release	Extreme	None	Very powerful	Moderate

- Atom dissolution: All atomic elements and quarks dissolve and release their energy.
- Fusion: The atomic elements fuse into a new, large atom.
- Fission 1: Some atomic elements are knocked loose, dissolve, and release their energy. The atom is able to rebuild the missing elements and continues as an atom with slightly less energy.
- Fission 2: No atomic elements are knocked away from the nuclei. The uni particles' repulsive force can protect the nuclei, but the increased repulsive force releases energy to the environment (friction).

Fusion Power Plants

Over the last few decades, many attempts have been made to create energy using fusion on Earth. Almost none of these have produced more energy than has been used to facilitate the fusion process. Many different designs have been tried for such a power plant, and focus is currently on a tokamak design and the use of two hydrogen isotopes, deuterium and tritium.

Hydrogen normally consists of one proton and no neutrons. Deuterium is an isotope of hydrogen consisting of one proton and one neutron. The isotope tritium consists of one proton and two neutrons. These isotopes are rare, and it takes a lot of time and energy to produce them.

The fusion process of the hydrogen isotopes deuterium and tritium then becomes:

(PN) + (PNN) -> fusion process -> PPHH (helium) + N
P = Proton
N = Neutron
PN= Deuterium (isotope of hydrogen)
PNN= Tritium (isotope of hydrogen)

In the fusion process, deuterium and tritium are fused into helium. During the process, one neutron becomes redundant and is released, dissolves, and its energy is released. Energy from this neutron can be converted into electrical energy by heating water into steam. Steam can be used to produce energy using ordinary generators.

Fusion Process and Energy

A lot of energy is required for the elements to fuse. They must be supplied with a great deal of energy so that the paths of the uniparticles of different atoms become wide enough for them to cross each other.

If uniparticle trajectories cross each other and atomic nuclei touch, they must preferably be moving in the same direction for them to fuse. If they collide at too sharp an angle, they can crush each other. We will then get an atomic dissolution reaction where nuclear elements and uniparticles are dissolved, energy is released, and we get free neutrons. However, if the speed is moderate, we can still create a fusion without the atoms needing to move in the same direction.

A great deal of energy is required to produce the isotopes deuterium and tritium, more in fact than a nuclear fusion process can release.

Therefore, a fusion power plant cannot produce an energy surplus. In addition, there may be a risk of an uncontrolled nuclear dissolution process occurring, with very unfortunate consequences.

Lawrence Livermore National Laboratory – Fusion

On December 5, 2022, a fusion experiment was conducted at Lawrence Livermore National Laboratory (LLNL) where more energy reportedly came out of the fusion material than was sent in. Powerful lasers irradiated hydrogen isotopes (deuterium and tritium), which released 50% more energy than the lasers sent into the isotopes. The isotopes contained one and two additional neutrons, respectively, compared to ordinary hydrogen.

This was interpreted as the start of a fusion process (fusion ignition), where the hydrogen isotopes fused into helium.

Fusion between common elements does not provide energy according to the Ilefos model. However, fusion between individual **isotopes** can produce energy. We interpret this as a fusion process where the isotopes were quickly supplied with an extremely large amount of energy. This energy gives the uni particles wide trajectories, where the isotopes' uni particle orbits cross each other and atomic nuclei collide.

The atoms did not move in the same direction, and we did not get a soft collision, but a more powerful collision. This collision between the atomic nuclei was not powerful enough for us to get a complete atomic dissolution, but it was most likely powerful enough for the isotopes to fuse, and an excess neutron to be released. The collision then brought the atomic nuclei into contact without shattering them. Two atomic nuclei that touch each other without shattering will fuse. The free neutron after fusion did not have enough speed to collide and release new neutrons from external atoms, thus creating a fission or atom dissolution process. After a short time, the neutron was dissolved and released the energy in the neutron's quarks. An atomic element without an atomic quark loses its internal management and dissolves rapidly if another atom does not absorb it. In this case, the neutron is not absorbed and is therefore dissolved.

The neutron did not have a high enough speed for a nuclear dissolution reaction to be started. When the free neutron hits the uni particles of other atoms, it is either broken into pieces or it dissolves itself since it does not have access to an atomic quark (control quark).

The hydrogen isotopes initially had neutrons. Two tritium atoms that fuse give two excess neutrons. When tritium and deuterium fuse, this results in one excess neutron. If two deuterium atoms fuse, no excess neutrons are produced. If two ordinary hydrogen atoms or an ordinary hydrogen atom and deuterium fuse, atoms must use internal energy to create neutron(s), establishing an atom with less energy in the atomic elements than the atoms that started the process.

Neutron and energy production by fusion of hydrogen and hydrogen isotopes into helium can be summarized as follows:

Table: Comparative Overview of Atomic Interaction Mechanisms

This table summarizes key differences between atomic fusion and fission, as well as other forms of nuclear transformation, according to the ilefos model. It highlights the role of energy, mass, track formation, and structural reorganization in each process.

Fusion between:		
Tritium and tritium	+ 2 neutrons	Produces energy
Tritium and deuterium	+ 1 neutron	Produces energy
Deuterium and deuterium	0	Neutral
Deuterium and hydrogen	- 1 neutron	Reduced energy in elements
Hydrogen and hydrogen	- 2 neutrons	Reduced energy in elements

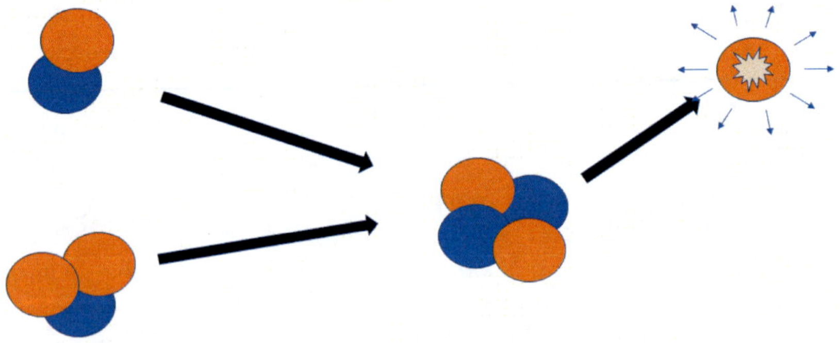

Figure 51. Fusion between the hydrogen isotopes tritium and deuterium to form helium. The process produces one free neutron, which dissolves (Bent R Pettersen).

In fusion, a soft collision occurs where the number of nuclear elements (protons and neutrons) is retained.

In this fusion process with the hydrogen isotopes, energy from the excess neutron is released. The isotopes had more neutrons than ordinary hydrogen, and tritium, which has two neutrons, also has two uni particles. When tritium loses a neutron, it also loses a uni particle. This uni particle is dissolved, and the energy is released.

This process happens quickly, resulting in a fast release of energy. Energy is produced from dissolved neutrons and uni particles due to a fusion process between the isotopes. In the fusion of common elements that do not have extra neutrons, we will not get an extra nuclear element (a neutron) after fusion. We do not, therefore, end up with the dissolution of an excess nuclear element, which releases energy.

Normal fusion -> does not give energy

Fusion of isotopes of matter that have an extra atomic element (neutron), gives extra atomic element(s) after fusion, which dissolves -> producing energy

The experiment at the Lawrence Livermore National Laboratory was remarkably interesting, but it will take a long time before commercial energy production is possible using this method. The project used 192 lasers to start the fusion, consuming 300 megajoules of energy. They delivered 2 megajoules to the fusion isotopes. The isotopes released 3 megajoules of energy. The process provided energy in a fraction of a second. The lasers must be cooled for several days before they can be used again. It is also unclear how to deliver a series of isotopes into the reactor. The reactor walls become radioactive in the process, creating radioactive materials. (When an atomic nuclear element dissolves, the dimensional quarks also release their energy. Excess dimensional energy creates radioactive radiation. Tritium is an expensive isotope. One gram costs about 30,000 USD, making it one of the most expensive substances on earth.

Lawrence Livermore National Laboratory has an annual budget of nearly 2.8 billion USD. A commercial power plant will probably have a budget of more than 3 billion USD. The process is complicated and is associated with great uncertainty. It is therefore unknown whether the process can compete in terms of price with existing technologies for energy production.

For a nuclear reaction to produce energy, one of the following must occur:

1) An atom is drained of energy
2) An atomic element dissolves

In order for an atom to be drained of energy (event 1), a fission process must take place. Here, atomic nuclear elements (neutrons) and uni particles can dissolve and release the energy, but the element/particle is rebuilt immediately and the atom remains unchanged in composition.

If a partial nuclear dissolution occurs (event 2), all the energy in the nuclear element/particle is released, and this is not recreated. We get an atom with fewer nuclear elements; in the case of total atomic dissolution, the atom is completely dissolved, releases all energy, and ceases to exist.

In normal fusion, all nuclear elements in the fused atoms are retained in the new atom after fusion. No nuclear element is dissolved or energetically depleted. A normal fusion does not, therefore, produce energy.

In a fusion between atoms with an extra nuclear element, such as one or two extra neutrons, one or more extra neutrons are left over after fusion. In the fusion of hydrogen isotopes, the new element has two protons and two neutrons (Helium), and the neutron(s) that become redundant are released and dissolved. Energy from the neutron(s) that are dissolved is released, thereby producing energy.

Bent Rolf Pettersen, 2022

Chapter 52

Conclusion

A Unified Energy-Based Explanation of the Universe

This book has presented a new explanation of physics, based on the properties of different energies and their interactions, which offers a coherent account of universal dynamics.

The key elements of this new explanation of physics based on the Ilefos model are as follows:

Each type of energy exhibits distinct properties. Physics is how pulses of energies react with similar or different energies. A force is a reaction between different energies. Energies create forces based on their properties and their interaction with each other.

The Ilefos model is a qualitative alternative conceptual model based on the properties of energies and how they interact.

Ilefos are energy units, pulses of energy, which have an attractive force toward each other. When pulses of this energy are released, the pulses travel at a release speed. The attractive force between the energy units makes them form a track of ilefos energy units, i.e., a gravity track. The energy units travel at a speed that increases with distance from the source. This creates greater distances between the ilefos energy units, thus weakening the ilefos track with distance from the source.

The attractive force of an ilefos track attracts other external ilefos tracks. When different tracks meet, they form a connection. Two strong ilefos tracks form a strong connection, as seen in atomic bonds. When two weaker ilefos tracks meet further out, they form a weaker connection in the form of gravity. Eventually, the distance between the ilefos energy units in the track becomes too great, and the attractive force can no longer keep them together in a track. The track then dissolves into free ilefos energy units not connected to a track or source. This is dark energy.

Atoms release ilefos energy units, which form ilefos tracks (gravity tracks). To be in energy balance, they must constantly receive the same amount of energy from dark energy, free energy units in the vicinity. Atoms and universal phenomena then recycle energy units. This explanation of the attractive force seems to be applicable in both quantum mechanics and astrophysics.

Ilefos is the attractive energy that has given the name to the Ilefos model. Other energies have other properties.

Energies can be concentrated in quarks and organized into energy systems. Together, the reactions between the different energies seem to be able to explain the dynamics of our universe.

This model can explain dark matter and other universal dynamics, such as the accelerating expansion of the universe and the Big Bang. The Ilefos model thus appears applicable across both quantum mechanics and astrophysics.

Chapter 53

Closing Remarks

Despite decades of intense research, many of the most fundamental questions in nature remain unanswered, such as: What is gravity? What is dark energy and dark matter?

Current theories often rely on hypothetical constructs to remain mathematically consistent and often lack a clear cause-and-effect explanation. A correct theory of physics must work with quantum mechanics and astrophysics. We have several approaches to this, but none seem to give a clear, universal cause-and-effect explanation.

Why have we been unable to find answers to the basic questions of nature and reconcile quantum mechanics and astrophysics? Could we be confined within a scientific framework that inherently limits our ability to answer such questions? Are we working within a research paradigm that cannot provide solutions? Could the lack of answers be an indication that we need a paradigmatic debate? Do we need a paradigm shift within research?

The Ilefos model is a proposed new physics model that seems applicable within both quantum mechanics and astrophysics. It also seems to be able to explain dark energy, dark matter, and universal dynamics like matter, stars, black holes, and the accelerating expansion of the universe.

An extended version of the Ilefos model can also explain big bangs and the grand universe outside our visible universe.

This qualitative alternative conceptual model may be used to form the basis of a new understanding of physics. For this to work, we need more advanced mathematical models based on the theory. I am currently developing such models, and I encourage others to xplore and expand upon this foundation.

The visualization of components in atoms like quarks, protons, neutrons, and uni particles are for illustration purposes. We do not know how they are constructed, but I have devised these visualizations based on the atomic dynamics of the Ilefos model.

I hope this model may contribute to a deeper understanding of nature and the architecture of our universe.

Bent Rolf Pettersen, 2024.

Index

N

P

Q

R

S

T

U

V

W

Y